WORKING BEYOND BORDERS

APPLYING GIS

WORKING

BEYOND

BORDERS

GIS FOR GEOSPATIAL COLLABORATION

Edited by
Jill Saligoe-Simmel, PhD
Maria Jordan

Esri Press
REDLANDS | CALIFORNIA

Esri Press, 380 New York Street, Redlands, California 92373-8100
Copyright © 2023 Esri
All rights reserved.
Printed in the United States of America.

ISBN: 9781589487628
Library of Congress Control Number: 2023939041

CONTENTS

PART 3: ENGAGING COMMUNITIES   61

PART 4: BUILDING CAPACITY   95

NEXT STEPS   126

INTRODUCTION

COLLABORATION IS A FUNDAMENTAL TENET FOR MOST organizations. A collaborative environment builds trust and confidence in how work gets done and helps staff work better together to achieve shared goals. In today's world, collaboration extends far beyond the boundaries of an organization's structure or jurisdiction. Governments, businesses, academia, and nonprofits are breaking through barriers that limit effective data sharing and collaboration between people who are often addressing similar issues.

Many organizations use geographic information system (GIS) technology to understand and solve problems, ranging from the local to the global scale. Geospatial thinking is a mainstay for decision-making in today's world. Maps and geospatial analytics offer potent ways to understand, manage, and solve complex problems. Location helps answer not only questions such as where things are, but what they are, how they relate to each other, what they mean, and how they change over time. Geography creates a familiar landscape where people can visualize issues to cooperate regardless of perspective or objective. GIS is anchored to an underlying infrastructure designed for collaboration.

Geospatial infrastructure is a *system* that connects people, processes, data, and technology. At its core, a geospatial infrastructure adheres to the fundamental principles of Web GIS, taking advantage of the internet and cloud computing to share data and collaborate through an interconnected network of systems and portals.

When organizations integrate their geospatial infrastructures, a *system of systems* emerges that enables them to interconnect and work across borders, jurisdictions, and sectors. Integrated geospatial infrastructure is an emerging implementation pattern that melds spatial data, geospatial technologies, and supporting systems and processes to enable informed decision-making across industries and levels of government. Today, organizations are forming alliances and changing the rules of collaboration by incorporating geospatial engagement through mapping and GIS applications.

With geospatial infrastructure, organizations can better manage geospatial data cooperatively and create online destinations, such as hubs and open data portals, in which partners and other organizations can participate, contribute, and benefit from shared knowledge and authoritative information.

This book provides a glimpse into the ways that governments, businesses, nonprofits, and others are building collaborative environments using GIS. The book is divided into four parts: governing collaboratively, providing decision-ready data and technology, engaging communities, and building capacity. Each section suggests the next steps readers can take to investigate, create, and use geospatial infrastructure for improving collaboration.

Governing collaboratively

Part 1 introduces integrated geospatial infrastructure as a technology enabler containing components that build trust and ensure reliable communication among partners. Real-life stories show how organizations approach collaborative governance by accommodating an adaptive and interconnected web of people, processes, data, and technology. Collaborative organizations are responsive to their communities of interest and representative of their participating partners. Collaborative organizations work together to define elements of leadership, vision, strategies, policies, and reporting on performance indicators.

Providing decision-ready data and technology

The stories in part 2 demonstrate how organizations provide geospatial data that is findable, accessible, interoperable, and reusable to support decision-makers, users of different disciplines, and partner organizations. This section introduces the concepts of geospatial hubs, portals, and applications that drive collaboration. Decision-ready data and technology elevate the value of geospatial data and support breaking down administrative barriers and eliminating data silos, giving users access to the information they need.

Engaging communities

The stories in part 3 describe how collaborative organizations recognize the need to empower their audiences with knowledge and understanding through geospatial information and technology. With a geospatial infrastructure in place, decision-making improves, and collaboration becomes the standard behavior between people and organizations. GIS, as a system of engagement, enables everyone to get the information they need, making collaboration across organizations possible.

Building capacity

Part 4 demonstrates how the promise of geospatial collaboration continues beyond the implementation of GIS, making geospatial data, maps, apps, tools, and solutions more accessible to individuals and organizations that might not have the resources or expertise to access them on their own. Geospatial collaboratives offer training programs from basic concepts to advanced techniques, including workshops, online courses, and conferences. Additionally, these collaboratives provide mentorship opportunities to young professionals, developers, students, and staff, supporting innovation and creativity within the geospatial community.

HOW TO USE THIS BOOK

THIS BOOK IS DESIGNED TO HELP YOU ADD GEOSPATIAL thinking to decision processes and improve collaboration. It is a guide for taking the first steps toward integrating geospatial infrastructure among organizations and applying location intelligence to common problems. Using the information in this book can help create a more collaborative environment between you and your department, your partners, and partnering agencies and organizations. You can use this book to identify where shared data, maps, spatial analysis, and GIS apps can support your work and then, as a next step, learn more about those resources and strategies.

The concluding section of this book offers a basic strategy for taking the next steps in applying GIS collaboratively in your community of interest.

Learn about additional GIS resources for geospatial collaboration by visiting the web page for this book:

go.esri.com/wbb-resources

The Minch
Moray Firth
Aberdeen
Dundee
Edinburgh
Glasgow
North Channel
Belfast
Newcastle upon Tyne
Douglas
Irish Sea
Preston
Leeds
Dublin
Liverpool
Manchester
Sheffield
Stoke-on-Trent
Nottingham
Leicester
Wolverhampton
Birmingham
Coventry
St George's Channel

PART 1

GOVERNING COLLABORATIVELY

ORGANIZATIONS NEED COHESIVE, RELIABLE SYSTEMS and people to operate efficiently and effectively. To do this, organizations enact different forms of governance for employees, assets, and daily interactions between business units, offices, and leadership. In simple terms, governance is a formal approach that helps establish who is responsible for what and how decisions are made. Governance sets direction by establishing a strategy and addressing leadership, authority, decision-making, accountability, and a holistic business alignment approach. Governance helps successful organizations achieve goals, maintain consistency, and ensure that decisions are on target and benefit everyone involved.

Similarly, governance provides a structure for collaborative organizations. Experience worldwide shows that thriving cooperative alliances require a few main components. These components include an integrated approach, data stewardship, and an organizational structure representative of the participating partners. This approach entails setting direction through geospatial strategies, policies, and practices and measuring progress on key performance indicators (KPIs).

The following example illustrates how collaborative GIS provides a region with a strong foundational information system.

Through governance, partners gain a more complete understanding of adjacent and overlapping responsibilities and where they can work together to address common goals. In this example, the city transportation department uses GIS to map streets, monitor traffic and accidents, and deploy maintenance crews. Other departments need that data, too. The public safety department manages addresses and uses the same street network so they can be aware of street closures, route emergency vehicles, and have situational awareness. The public works department uses the transportation network to plan installation of new sewer lines. The local health department uses roads, buildings, addresses, and utility data to inspect local businesses and shares its findings with the state health department and the public. Regionally, planning authorities use this data from the city, combined with similar data from surrounding cities and towns, academic research, and statewide and national data, to develop long-range regional plans for economic development, improved quality of life, and sustainability.

Although this example involves local and regional partners, geospatial collaboration occurs on the local to global scale—from whole-of-government initiatives to modern spatial data infrastructure (SDI) initiatives.

Governance in a collaborative alliance

- creates a system of accountability that defines and enforces the rights and responsibilities of stakeholders,

- builds trust and reduces friction by providing transparency,

- decides what components of the geospatial infrastructure each partner is responsible for, and

- agrees on how partners apply rules and standards to ensure an efficient and effective operation of the system of systems.

By sharing location intelligence in an integrated geospatial infrastructure, partner organizations can work collaboratively and in concert to coordinate their plans and activities. With GIS, organizations can maximize the value of their data while adhering to governance direction, policies, and supporting jurisdictional priorities. Governing collaboratively facilitates a more balanced and proficient approach to working together.

Organizations use collaborative governance to

- drive geospatial strategy aligned with broader mission areas,

- enable a sustainable, flexible, and purposeful GIS technology architecture,

- establish data governance, sharing, stewardship, and quality to meet the end-users' consumer requirements,

- build workforce capacity,

- deliver access to GIS content through hubs and open data portals to engage with stakeholders, and

- prioritize investment decisions and monitor progress.

Real-life stories

The real-life stories in this section present examples of how organizations use location intelligence and GIS to establish collaborative governance within their organizations and between other agencies, outside partners, and the public.

DATA COOPERATIVE ENABLES COLLABORATIVE TRANSFORMATION IN YORK REGION

Esri Canada

THE YORK DATA CO-OP IS THE LATEST AND PERHAPS THE most compelling innovation of the YorkInfo Partnership. Since 1996, the partnership of the Regional Municipality of York (York Region) and its nine municipalities, two district school boards, and two conservation authorities has been recognized for its culture of cooperation that enables the partners to collectively benefit from location intelligence.

The York Data Co-op connects the partners in the YorkInfo Partnership in a distributed environment and gives them access to each other's digital assets (data, applications, and tools). ArcGIS® technology provides a straightforward and secure enterprise platform making it possible.

"Enabling data access and sharing has been our core direction for years," said John Houweling, director of the Data, Analytics, and Visualization Services Branch at York Region. "Building trust in the data adds confidence in decision-making and results in more effective services. The York Data Co-op takes this further by enabling partners to share apps and tools as well; increase collaboration in common business areas; and in so doing, provide a foundation for change."

As Houweling noted, the partners also use each other's GIS scripts, code, and apps so they can build and endorse common data models, realize efficiencies, reduce risk, and deploy new spatial applications that support staff. Not every partner has the internal capability to create these tools, so accessing and using these GIS assets creates immediate impact and benefit.

The project also facilitates a federated Open Data Portal that allows York Region residents and businesses to access government data. As the partners' open data sites are connected, the data exposed by any partner on their own open data site becomes seamlessly discoverable and available on each partner's site.

Participants publish their digital assets by registering them within the co-op. The asset is then available for others to find and use. Assets aren't physically copied into a central server but remain managed and stored behind the firewall of each partner's domain. This way, the assets are "live" and always current, and users do not need to deal with versions on different servers. As a federated solution, no one partner is the main distributor or owner of the collective assets; each organization maintains autonomy over its own digital assets. The York Data Co-op is creating a sole source for information.

For example, a search for "street trees" will return all information sources for this type of data, apps, and code, regardless of where they're stored. The system knows where to find the digital assets and displays all results in a consolidated report on-screen.

Other digital assets provide insight and added value and may meet an immediate business need. These assets include interactive maps, dashboards and information products, development tools, APIs, and operational business apps. By using the apps and services that others have developed and shared, the partners can build their own capacity and capability and learn through sharing. By using shared digital assets, they can deliver solutions that they may not otherwise have had the time or ability to undertake.

The York Data Co-op acts like a virtual marketplace and provides a familiar online shopping experience for partners to search for and acquire published digital assets.

Digital assets are grouped based on their function and use, such as property services, infrastructure, environment, and health, rather than on their origin. Users find assets in the co-op in the same way a shopper finds groceries at the supermarket—by type, not origin. As a result, users can scan through and download the latest digital assets quickly without having to trawl through assets unrelated to their search.

Each digital asset has a description (metadata) that includes its source, extent, timeliness, and other information. Metadata supports the cataloging and search functionality and provides information that the user needs to understand the asset before downloading or using it.

The co-op helped facilitate the City of Vaughan's Emergency Management Portal. Led by Vaughan Fire and Rescue, the portal uses the co-op and gives emergency management services staff from Vaughan, York Region Emergency Management, and York Region Police the ability to share current situational awareness data, such as incident area boundaries and road closures. This capability enables them to better communicate, as needed, with the public. Saving time during an emergency is just one example of how the portal proves its value.

Increasingly, the partners' programs realize how the co-op can be an effective infrastructure for the collaborative delivery of municipal services within and across the region. Services already identified as targets for a common solution include emergency management, municipal comprehensive review, building permits, construction projects, digital plan uploads, open data, zoning, sidewalks and streetlights, trails and parks, street tree management prioritization, and road collisions.

The York Data Co-op uses ArcGIS technology, including ArcGIS Online, a cloud-based mapping and spatial analytics service that supports a federated architecture and provides partners with a low-cost entry point and the ability to develop solutions.

"The ArcGIS Online development environment is so straightforward that we've literally been working things out on a whiteboard in the morning and then looking at a prototype online that afternoon," said Brendan Coles, a GIS project specialist for the Regional Municipality of York. "Everyone gets excited when they see how quickly we can make things real."

The partners build the federated platform using ArcGIS Hub℠, the same technology that underpins York Region's Open Data Portal, providing the virtual marketplace and shopping cart-like experience.

A version of this story titled "Data Co-operative Enables Collaborative Transformation in York Region" by Chris North originally appeared in the *Esri Canada Blog* on October 2, 2019.

IRELAND PROVIDES ITS RESIDENTS A GREATER SENSE OF PLACE

Esri

NATIONAL MAPPING AGENCIES, LIKE MANY OTHER industries, have felt the impact of internet disruptions. These disruptions in the flow of information have caused people and organizations to rethink what they do and how they do it.

Ordnance Survey Ireland (OSi), which merged with the Property Registration Authority and Valuation office to become a new state agency called National Geospatial ICT in the National Mapping Division, Tailte Éireann, in 2023, transformed its operations with a focus on adding intelligence to its data. This step allows OSi to offer new products and services and to stream data as an online service to those who want the most accurate and current maps for answering location questions.

Before the merger, OSi already had changed from a product business (making maps) to a services business (delivering location intelligence).

Conveying this change to stakeholders—including the general public, government partners, and businesses—began by adding attributes to spatial data about the *what* and *why* of map data, which is much more than what can be seen on a map. OSi shared the capability of this new intelligent data by replacing a static online map-viewing website with a collaborative national mapping platform.

In November 2015, OSi launched GeoHive, a modernized online system in which users can access map data from OSi along with 439 layers of map data from 29 data providers.

"GeoHive brought it all together," said Tony Murphy, former business and marketing manager, OSi. "You can go to a one-stop-shop to access data, join the layers, and visualize them."

Before GeoHive, a user had to contact each individual data pro-
vider to gather details. The online system allows users of the free
platform to combine, view, and query map data online. GeoHive lets
users analyze and interpret map data to understand relationships,
patterns, and trends for a more complete understanding of place.

Users of the system can access an array of information in the
combined catalog:

- **Education data,** including school locations and the
  education level of residents

- **Energy data,** including geothermal and wind energy sources

- **Environment data,** including air and water pollution and
  sensitive areas

- **Housing data,** including type, availability, and price

- **Land data,** including a national soil map, woodlands,
  greenbelts, and land use

- **Planning data,** including local area plans, historical areas,
  and land zoning

- **Population data,** including age, gender, and
  employment skills

- **Real estate data,** including architectural heritage, housing
  starts, and development plans

- **Safety data,** including the location of fire and police stations
  as well as crime maps

- **Transportation data,** including airports, roads, trails,
  parking, and scenic routes

- **Services data,** including library locations, arts, playgrounds,
  sports facilities, and public toilets

- **Water data,** including historic floods, river widths, water flow, and water levels

GeoHive also provides the ability to create custom maps online. Users can combine the layers of different information, make maps, save them, print them, and share them through social media. This ability to join or "mash up" the information on a map gives users the means to collect and display information relevant to their lives.

"We visualize wider stakeholder engagement and the creation of a *hub of hubs*," said Lorraine McNerney, former general manager of OSi and general manager of National Geospatial ICT. "It's about communicating policy, informing the public, and measuring with an eye on progress."

Thanks to OSi's linking with partners, the people of Ireland gained free access to map data that includes current and historical maps, details on the land such as geology and hydrology, and information about the populace.

Mapmaking through GeoHive supports residents and businesses. A residential property viewer, for example, can allow prospective buyers to gather data from sources about the conditions around a site, such as transportation, education, employment, hospitals, and crime. This information can be compared for multiple locations to help potential homeowners make informed decisions on where to buy. And business owners, for example, can use a similar set of tools to help site new offices and retail outlets. They also can view information about the population, job skills, infrastructure (electric, communications, and water utilities), and transportation to determine the site best suited to their operations.

Through the National Mapping Agreement, all government bodies in Ireland gained access to OSi's geospatial data since January 2017. In Great Britain, thousands of organizations have taken advantage of a similar program run by Ordnance Survey Great Britain since 2011.

Great Britain has benefitted from improved data sharing in many ways:

- London has achieved better coordination of road closures for utility excavations and building construction. A new works planning tool uses accurate spatial data from many departments to analyze the impact of each project and identify potential conflicts with other work planned in the same area.

- England's Bolton Council used transport and topography data to better plan waste collection rounds to 123,000 residential properties. Designing optimum routes designed to reduce fuel and vehicle maintenance costs has saved £400,000 per year.

- Improved data for the Isle of Wight has been credited with improved ambulance service on the island, allowing crews to quickly locate a patient on a map and help the ambulance service meet its goal of an eight-minute response time.

The availability of consistent and accurate data improves the efficiency of data maintenance and delivery, leading to time and cost savings with less administration required. This, in turn, aids analysis and decision-making across government departments and national, regional, and local governments.

OSi behind college-level mapping and data services

College-level educational institutions also have made use of OSi mapping services and data for their projects and research. For example, the academic research project Programmable City accessed the data to investigate how cities are increasingly being translated into code and data and how this change impacts the lives of people in their cities.

The GeoHive online mapping platform can be partitioned into microsites for individual groups, showing the most relevant data.

"We envision setting up groups for education and local government for collaboration," Murphy said. "Doing this will provide integrated access to information for a wide range of users and support the implementation of public policy and initiatives."

GeoHive started as a replacement map viewer, adding a broad catalog of authoritative spatial data, and the ability to make a map. It now provides a backbone that the community is building on.

A version of this story titled "Ireland Provides All Citizens with a Greater Sense of Place" by Mark Cygan originally appeared in the *Esri Blog* on October 20, 2017.

OPEN DATA GIVES LOS ANGELES A BOOST IN COMMUNITY COLLABORATION

Esri and the City of Los Angeles

A PIONEER OF OPEN DATA, THE CITY OF LOS ANGELES WAS the first city to launch a site dedicated to exploring, visualizing, and downloading location-based open data via the city's GeoHub. Eva Pereira, LA's chief data officer, outlined some of the new areas of focus in this question-and-answer format. Some of these initiatives focus on equity and social justice, with geospatial analysis playing a leading role in addressing the city's challenges.

Q: Los Angeles is well recognized as a leading data-driven city. What role does the yearly Open Data Day play?
A: We try to do something new every Open Data Day. On March 5, 2022, we hosted a citywide competition to improve the quantity and quality of our open data. We had city departments look at their open data and improve their descriptions, improve their metadata, and in some cases publish new datasets.

In 2021, we focused the event on the pressing issue of food insecurity in Los Angeles, which was heightened by the pandemic. We partnered with researchers from the University of Southern California's Keck School of Medicine, the LA Food Policy Council, and LA County to talk about the topic. Then we had a hackathon where we shared the data and participants developed solutions.

Q: How do you use open data to engage with community stakeholders?
A: I try to get partners involved that are mission aligned. There are problems that the city is trying to tackle, and we have our data, but there are community-based organizations, research institutions,

nonprofits, advocacy groups that all are trying to solve the same problems.

Q: What are some issues the city addresses using a geographic approach?

A: For everything from land-use planning to transportation and broadband access, we use geographic data to pinpoint areas of need. For the 2022 Open Data Day, we gave the Department of City Planning the Department Data Champion award for updating the data they share publicly through the GeoHub. They share a lot of interesting map-based data around issues such as enhanced bicycle networks. The department has a huge repository of data and maps that they keep fresh, related to zoning, pedestrians, vehicles, and transit.

If we maintain our open data, then we're able to dive into these issue-specific problems and tackle them effectively. It starts with departments maintaining what they've posted, publishing new data, and continuing the commitment to open data. Future initiatives are built upon open data.

Q: What are you doing in the area of examining equity?

A: We're working on an equity site that explores the history of racial injustice in Los Angeles and connects residents to information and resources related to city services. We've divided it by issue, such as housing and built environment, economic opportunity, education access and attainment, justice and policing, and health and well-being. It's going to be a great information tool to understand our history and how we got to where we are today. I'm hopeful that residents will use it to connect with resources and get involved.

For another project on health equity, we've used CalEnviro-Screen 4.0 data to visualize pollution and health impacts for LA neighborhoods. The city is a partner on a NASA air quality grant.

We mapped the location of all the sensors maintained by the Bureau of Streetlights to identify where we need to place more sensors to capture more accurate air quality data. We're very committed to environmental justice, and that's a problem that data can help solve.

Q: We've seen a great increase in the use of dashboards to track progress. Has that been a trend for you, too?

A: Dashboards are particularly helpful when you want to provide an information tool to policy leaders. We produce a lot of operational dashboards to let them know how a particular issue is trending. We need to understand the baseline and build data collection strategies to see whether we're on or off track. Dashboards help us keep an eye on things and see whether we're hitting the goals that we've set for ourselves.

We create a lot of story maps too, when we want to provide context around an issue, such as improving digital inclusion in Los Angeles, mapping Black-owned businesses and Paycheck Protection Program (PPP) loans, evaluating emergency rental assistance, and analyzing equity for capital improvement projects.

Q: How would you describe the role of data and maps for decision-making?

A: Data can help us improve city programs, services, and operations and inform policy change. We always start by defining the problem. A problem well-defined is a problem half-solved. And from there, we explore the available datasets and the outcomes we hope to accomplish by visualizing and diving into the data.

A lot of our problems can be solved by mapping to pinpoint the areas of the greatest need and then connecting those areas with resources. Our data projects typically fall into three buckets: resource allocation projects, gap analysis, and equity analysis projects.

The city is changing all the time, which is why it's important to have access to quality information and current mapping data. There are a lot of different problems that mapping can help solve.

A version of this story titled "Open Data Gives Los Angeles a Boost in Community Collaboration" by Nick O'Day and Eva Pereira originally appeared in the *Esri Blog* on May 10, 2022.

WHAT IS THE BUSINESS VALUE OF LOCATION DATA?

Esri

IF A FOOL KNOWS THE PRICE OF EVERYTHING AND THE value of nothing, as the saying goes, then double that when it comes to data. Any CIO footing the bill for enterprise technology knows how much it costs, but do they know the value of the data on which those systems run? Generally, no. For data-centric organizations, understanding the value of data may soon become an essential practice.

Less than 20 percent of business assets are tangible, and few firms have successfully valued the data that represents one of their greatest intangible assets. The traditional business balance sheet is partly to blame. There, physical assets such as buildings and vehicle fleets are valued explicitly, whereas data assets tend to be valued implicitly. But that scheme gives short shrift to the primary asset of the digital age. If data assets are in disrepair, a company might not give them due attention because their value isn't obvious. Data assets might not identify the consequence of ill repair on a balance sheet the way a failing HVAC unit or an underperforming store would.

To bring valuation practices in line with the times, executives are trying to put a value on their data. A pioneer in this movement is a government-owned company in the United Kingdom that established the monetary value of data.

Highways England is the UK's governmental organization charged with operating, maintaining, and improving England's 4,300 miles of motorways and major roads, which carry more than two-thirds of all freight. Highways England aims to connect the country and increase the safety and reliability of its major highways.

Leaders of the organization have brought together groups to understand and assign a monetary value to data and arrive at a common business language to justify data investments.

Highways England has assessed more than 60 datasets that create value for internal and external stakeholders, including any group that depends on road transit—logistics providers, retailers, manufacturers, major transport hubs, commuters, local authorities, and data consumers.

During the evaluation process, its analysts discovered that geospatial data about its road network was a particularly high-value dataset, covering everything from real estate, bridges, and work trucks to the places where weather events occurred and network roads.

"What we wanted to do is make data a bit more visible to the organization and to our people, our suppliers—from a perspective that they understand, which tended to be finance," said Davin Crowley-Sweet, the agency's chief data officer.

Crowley-Sweet first experimented to uncover the types of data stakeholders prioritized, as well as the financial value of that data. By gaining a mutual understanding of data's value, he reasoned, the organization could better structure its investments to protect, maintain, and build its data assets.

During the first stage of the assessment, Highways England valued its physical road infrastructure at £115 billion and the intangible value it delivered to the country at £200 billion, according to Crowley-Sweet. Next, to establish the data's value, a Highways England team called representatives from every stakeholder group—59 interviews in all. Its team talked to local authorities, major transport hubs, and data consumers and identified the datasets that most supported their activities.

The team supplemented its phone interviews with online surveys to ensure the process was statistically robust. About 300 people weighed in during the nine-month process.

"We asked those stakeholders to rank the Highways England initiatives they valued the most—things like maintaining the road surface and providing traffic flow data," said Victoria Williams, head of data and information governance. The results yielded a measurable preference for each initiative, which the team then used to gauge data dependency, identify enabling datasets, and calculate the economic value created for each stakeholder group.

The value of organizational data was estimated at almost £40 billion for the top six stakeholders, and about £60 billion for all groups combined. The value resulted from time saved through traffic notices and roadwork alerts, safety improvements, and reducing emissions and noise.

The team then posed this question: "Our governance structures are geared around managing physical assets, but did you know that our virtual assets are one-third of the value of our physical assets?" The inescapable conclusion: the organization was not paying enough attention to its data assets.

The insight that followed—embedded in these questions posed by the team—drew the attention of senior leaders:

- Did you know that our data is worth four times the value of the technology that houses the data?

- Are we investing in the right proportions to protect and grow that value?

- Are we only in the business of buying new hardware and software?

The team found that the data that helped the organization reduce operating costs was not the same data that drove customer value. For instance, the organization traditionally focused on data related to road infrastructure, such as how much it cost to build and maintain new roads. But the data most valuable to stakeholders was information related to traffic flow, traffic speed, and weather events.

"We've been able to have a much more customer-focused lens on what datasets we need to focus on and why, which has brought an external dimension to how we view things," Crowley-Sweet said.

For instance, road maintenance crews did most of their work overnight on the assumption that that traffic is lightest then. But the team found that the logistics sector makes deliveries at night so that shops have fresh goods in the morning. This insight allowed Highways England to adjust the timing of maintenance tasks.

"Doing this work has also got us to understand much more around what our stakeholders need from us rather than us thinking we're creating the value," Williams said.

The team established that, on average, a £1 investment in data by Highways England produces £2.7 in economic value for the logistics companies, commuters, transport hubs, and other groups that depend on its roads. Using that assessment, the team found that of the organization's 60 data types, geospatial data describing the location of the roads within the network was the most valuable, at £3.2 billion. Geospatial data helps workers effectively manage traffic, incidents, network capacity, safety, and the environment.

The next most valuable dataset involved traffic flow, at about £2.3 billion. Road information (accident alerts and other information pushed out to consumers) was valued at about £2.2 billion. For each type of location data, its intrinsic value grew when stakeholders—commuters, freight drivers, and others—made better decisions and improved efficiency and safety because of it.

Highways England repeats its data-valuation exercise about twice a year. For Crowley-Sweet, the effort to assign financial value to data and raise its profile pays off by increasing not just the return on data assets, but in the safety and well-being of stakeholders.

"We want to be sure that when you work with us, you get home to your families and loved ones safely," he said. "If we're sending people out to dig holes, I want them to know where the buried cables are. I want them to know where the ditches are, where the tripping hazards are."

Like other important data, that information is housed in a GIS that provides the information on smart maps.

The valuation exercise helped Highways England spot what businesses are eager to find—opportunities to bring datasets together and multiply their value.

Whereas a siloed business organization might ask questions such as, "Are we on budget?" and "Was the work done on time?" Highways England aims for insights such as how much it costs to run the M25 Motorway. Bringing data together enables decision-makers to answer questions about value instead of questions around cost.

"Location gives them a common reference point," Crowley-Sweet said. "But it's more than just being able to put things on a map—it's the ability to link different parts of the organization together to create the type of value that we're trying to create."

In Highways England's case, it means linking the world of assets with the world of finance. Any transaction—for instance, a road repair—incurs a cost at a location. "So you can start configuring your financial systems to contain elements of spatial data to function as a common denominator," Crowley-Sweet said. Then, rather than focusing on the cost of discrete transactions in a financial system, the organization can determine the cost to operate a service from location A to location B.

By seeking those patterns across datasets and anchoring financial data to location, Highways England can show a return on its data investments.

In addition to promoting data sharing, the project established a common language for discussing data and its value. "It was about being able to give people the right lexicon—standard terms the business would use to put together a business case for a new investment—so they could have a meaningful conversation and create shared [understanding] around what data means to them," Crowley-Sweet explains.

Valuing data within Highways England has focused the organization's management on its customers—its stakeholders. "We discovered that quite often, what we think is beneficial to the customer is actually not necessarily of true benefit to the customer," Williams said. Every such revelation has its origins in data, which helps establish and reinforce the data's value.

Highway England's stakeholders find value in understanding what is happening on the roadways, whether it's facilitating the delivery of food and perishable goods or knowing which roads are open and what condition they're in. These insights can have profound consequences on the survival of villages and communities, Crowley-Sweet said.

A version of this story titled "What Is the Business Value of Location Data?" by Marianna Kantor originally appeared in WhereNext on February 9, 2021.

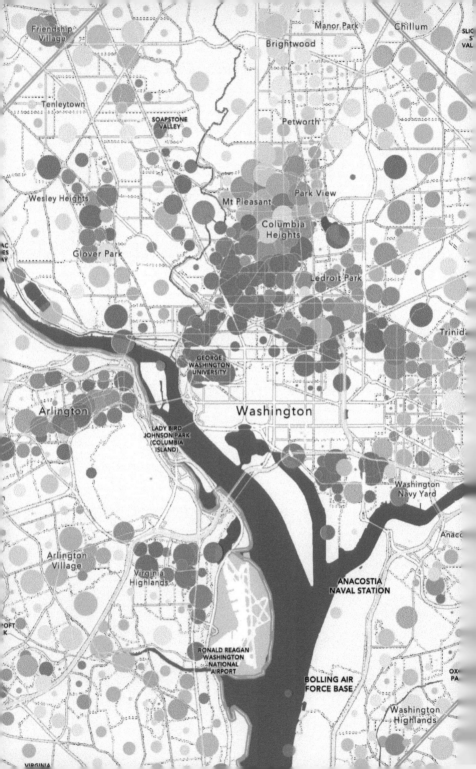

PART 2

PROVIDING DECISION-READY DATA AND TECHNOLOGY

GOVERNMENTS, PRIVATE COMPANIES, UNIVERSITIES, nonprofits, and NGOs collect enormous amounts of data. Organizations share and federate open and secure data worldwide to create an interconnected web of authoritative content at local, regional, and national levels. Ideally, they provide high-quality data that is findable, accessible, interoperable, and reusable (FAIR). In addition, organizations share foundation datasets that support a common operating picture for projects, programs, priority initiatives, policymaking, and scientific research.

Although these organizations strive to put all that data to the best possible use, they often must also present information to other people—decision-makers, for example. Therefore, data must be shaped into easily understandable formats that communicate the most relevant information and provide a statement about the issues that is thought-provoking and actionable. In other words, data must be decision-ready in all its shapes and forms so that it can be reused to tell a relatable story.

Beyond being FAIR, decision-ready data expands access to the broadest possible audiences, including knowledge workers, students, and the engaged public who may lack technical expertise but still require access to information the data reveals. Charts, graphs, and other infographics are traditional methods for punctuating important aspects of raw data. Still, they can lack the context that cements ideas and comparative statistics in the minds of decision-makers. Geography—where data happens, where it is derived from, and where it has significant relationships with other things—provides a common visual language, a contextual framework, and a knowledge infrastructure for understanding and action.

With GIS, organizations transform data into a source of location intelligence, helping decision-makers visualize and understand where trends are up or down; how widespread the issue is; and what nearby elements may be influencing program performance, policy success or failure, and any opportunities.

GIS allows users to model the real world geographically. As a result, they can create the basis for a digital twin of the city, the landscape, the environment, and the world. Planners, analysts, policy makers, and experts across disciplines can use digital twins to contribute, collaborate across organizational and jurisdictional boundaries, and share their work with partners, colleagues, and the public.

For example, the tiny island country of Grenada is confronting an uncertain future in the face of climate change. Needing a way to put vast quantities of data to use, Grenada used GIS to create a digital twin to visualize climate change challenges and develop solutions. The digital twin provides a complete view of the country and has a predictive capacity. It also has a foundational quality for reuse in countless other applications, such as planning the national census and conducting risk assessments. The data it comprises is now the basis for what the United Nations (UN) calls an Integrated Geospatial Information Framework (IGIF).

Expanding on the FAIR data principles, experience from successful organizations worldwide shows that geospatial data is decision-ready when it is

- easy to use, self-describing, interoperable, and licensed for reuse,

- easy to find, search engine optimized, and discoverable in a self-service global ecosystem,

- easy to understand, symbolized, and configured for quick visualization,

- relevant, analysis-ready data enabling access to quality features with well-documented attributes and fields for joining with other data,

- dependable, persistent, and optimized for scalability and performance, and

- accessible in various forms that serve broad audience requirements, including humans and machines, through APIs, downloads, and user-friendly applications for nontechnical users to explore.

Real-life stories

The real-life stories in this section explain how organizations use location data and integrated geospatial infrastructure to create shared authoritative data that effectively supports decision-making, improves discoverability, and sets the stage for engagement by stakeholders and the public.

CALIFORNIA CREATED A KNOWLEDGE BASE WITH GIS

Esri

CALIFORNIA — LIKE MANY OTHER STATES — COLLECTS DATA that increases understanding and, if integrated, leads to novel solutions. Often, however, this data is siloed in departments and not centrally accessible. The need to quickly locate the most current data on a topic is acutely felt during natural disasters, such as the wildfires that have cost lives and billions of dollars in damages in recent years. Departments scramble to locate the best data so they can effectively respond. In emergencies, having data available and accessible from a centralized portal can save a lot more than just time.

Working with Esri, the state created the California State Geoportal in 2019. The geoportal includes data and information products—web maps, apps, and storytelling maps—from dozens of state agencies. Many of the participating agencies had already created agency ArcGIS Hub sites to facilitate this sharing. Because the agency websites are federated through Hub, the most recent datasets from individual agencies are immediately available as the data is refreshed from each agency.

Since most data has a spatial component and these agencies used ArcGIS software, the portal allows state agencies access to integrate, analyze, and visualize the data using GIS. More than 40 of the state's departments use GIS, including the California Governor's Office of Emergency Services (Cal OES), California Department of Forestry and Fire Protection (CAL FIRE), California Census, California Department of Transportation (Caltrans), California Environmental Protection Agency (CalEPA), and the CalEPA Department of Toxic Substances Control (DTSC).

California created the Hub solution at no additional cost using Hub and ArcGIS Online.

Making publicly shared data and apps already developed by various California state agencies available from a single site helped immediately. By federating the agencies' websites using Hub, someone seeking state data no longer needs to know where the data originated to find it with relative ease.

Each dataset can be visualized using ArcGIS Online tools. California's residents can now visit just one site to answer a range of these and other questions:

- What is the sales tax in my area?
- What state parks are near me?
- What are the fishing regulations for a specific stream?

Implementing Hub as the solution to data access also provided different ways to visualize data, as well as tools such as ArcGIS StoryMaps℠ and ArcGIS Dashboards that require no GIS expertise. With this capability, the state makes data on water conservation, homelessness, wildfires, and other topics available and understandable.

This initial geoportal is part of a larger vision for facilitating data-driven policy development by extending the site with a nonpublic portion primarily for use by state analysts and researchers. Using Hub, analysts can retain and build on the results of their work.

The geoportal can have an even greater benefit when its data is combined with data from federal and local governments in analyses to inform and support more comprehensive government decision-making.

A version of this story titled "California Created a Knowledge Base with GIS" was originally published in the Winter 2020 issue of *ArcUser*.

DHS DATA HUB: OPEN DATA FOR ECONOMIC RESILIENCY

Esri

O N SEPTEMBER 11, 2001, A COORDINATED TERRORIST ATTACK on the United States devastated the country. At the time, the United States had little insight into the infrastructure and assets needed to guard against such attacks and respond to them.

One response was that the US Department of Homeland Security (DHS) and National Geospatial-Intelligence Agency (NGA) collaborated to develop the Homeland Security Infrastructure Program (HSIP). The federal agencies connected and aggregated data from hundreds of regional and local data providers to compile more than 500 national geospatial data assets and provide a more complete picture of our roads, water systems, schools, communities, facilities, and more. The result was the Homeland Infrastructure Foundation-Level Data Working Group, known as HIFLD.

HIFLD manages data aggregation, curation, and dissemination of data through computer discs. Access has been restricted for official use only (FOUO), limiting the access and reuse of this data to preapproved agency members.

The DHS mission takes an "all threats, all hazards" approach. But Homeland Security is more than just physical or emergency management. It must also address economic security and nontraditional, invisible threats to create a national resilience.

The DHS recognized the need to shift from its reliance on static media and embrace real-time, on-demand, and dynamic data that is available everywhere, by everyone.

The result is HIFLD Open—a hub that offers public access to more than 250 datasets as dynamic web services, downloadable files, and visualization tools for users to explore and use. For example, data about alternative fueling stations can help local governments evaluate transit infrastructure investments and fuel availability during a disaster event; data about shipping infrastructure indicates opportunities for new businesses that need access to goods transportation; and unique data for public refrigerated warehouses is important if there is a need to keep vital materials cold.

ArcGIS software provides a hosted infrastructure for DHS that provides data as web services and visualization and analytical capabilities to local governments and communities. Using GIS, DHS has curated and published the data across thematic groups so that anyone could explore and use it.

National security requires an integrated approach, one that HIFLD Open uses to encourage public participation. HIFLD Open

The Department of Homeland Security provides public access to more than 250 datasets.

has set a precedent to other governments across the federal, state, and local levels to also make more data freely accessible.

A version of this story titled "DHS HIFLD Open: Open Data for Economic Resiliency" originally appeared in Esri Insider on February 24, 2016.

WEAVING THE GEOSPATIAL FABRIC OF THE WORLD WITH AUTHORITATIVE DATA

Utah Geospatial Resource Center

I N OUR INCREASINGLY CONNECTED WORLD, COLLABORA-
tion can involve an infinite number of people. We can also contribute to products that go beyond the boundaries of our respective
organizations. An example in the public sector is how government
agencies collaborate to build statewide and national geospatial datasets. Collaboration, relationships, and agreed-on data standards hold
these datasets together.

Collaboration can be effective in the geospatial realm because
geospatial data is inherently visual and thus can be more intuitive.
Much like adding pieces to a puzzle, assembling data is rewarding.
GIS technology continues to make collaboration easier. The individual work previously performed and stored in silos can be integrated
to form a seamless, uniform geospatial fabric of the world.

Authoritative data can be defined as data that is provided by a
recognized source, such as a surveyor or governing entity (for example, an entity authorized to develop or manage the data, such as a
governmental jurisdiction). Placing authoritative data in the public
domain ensures that people can use the data.

Authoritative data is different from crowdsourced data, which is
typically generated by a large group of users contributing information at will and to the best of their knowledge. Crowdsourced data
can be useful for projects in which stewardship is of less concern—to
show live traffic data on a map, for example. It is also helpful when
information is needed quickly, such as during a humanitarian crisis
or after a natural disaster. In the midst of these events, individuals
must come together quickly and generate data without the additional
oversight that is often associated with authoritative data.

The clearest use cases for authoritative, open data are in systems such as 911, elections, health, and taxation. These systems support vital government services, such as public safety and emergency response; the creation, dissemination, and collection of election ballots; the monitoring of infectious disease outbreaks; and the setting of property and sales tax rates. In these cases, it is important to use authoritative data with a clearly defined and accountable steward.

Authoritative data is typically woven together through a process of aggregation. In this scenario, data stewards and data aggregators work with occasional overlap. Ideally, the data stewards work most closely with the assets. For datasets such as address points, land parcels, and road centerlines, the data stewards are often local government employees.

Data aggregators tend to play a facilitator role. When the aggregator is a government entity, it is usually positioned at the state or national level. In Utah, for example, the state's GIS office—the Utah Geospatial Resource Center (UGRC)—functions as the aggregator. The aggregator's role is to build relationships with data stewards, standardize the data, and produce a product that is diverse enough to be used in a range of systems and applications.

The steward-aggregator model has well-defined roles and contains accountability at all levels. One of the many benefits of this is that when an asset, such as an address point, is created or modified, it is done at the steward, or local, level. Through the process of aggregation, this data then flows into a larger and more encompassing dataset, such as springwater flowing from a high mountain stream into larger rivers, lakes, and, eventually, oceans.

This workflow reduces duplicate data and promotes a sole source of reliable information. It also ensures that users from the local to the national levels are using the same data. The more the data is used, the better it becomes.

Collaboration, relationships, and agreed-on standards hold the geospatial fabric together. Maintaining standards requires clear communication among stakeholders so that everyone understands their roles and each other's goals.

Data stewards typically create and maintain data for a specific purpose. However, when stewards, aggregators, and other stakeholders communicate, they often discover that they share goals and standards. Incorporating these goals and standards into workflows—by establishing naming conventions, for example, or using the same spatial reference system—makes data aggregation faster and more efficient, especially for statewide and national datasets.

Geospatial advisory groups are a way for various organizations to communicate their goals and solidify common standards. A few examples of these include the Federal Geographic Data Committee (FGDC), the National Geospatial Advisory Committee (NGAC), the Open Geospatial Consortium (OGC), and state-led advisory groups such as the Utah Geographic Information Systems Advisory Committee (GISAC).

The FGDC and NGAC are federally focused committees that work in conjunction with each other. Collectively, they seek to advance a National Spatial Data Infrastructure (NSDI) and develop and implement policies, best practices, metadata, and standards relating to geospatial data.

The OGC is an international consortium that adheres to the principles of making geospatial data findable, accessible, interoperable, and reusable, or FAIR. Its members create royalty-free, publicly available, open geospatial standards.

State-led advisory groups such as GISAC typically work to develop and update standards and best practices within their states. Clearly defining these standards is key to making speedy data updates within statewide datasets. In Utah, the Utah Geospatial

Resource Center, or UGRC, is able to aggregate the state's road centerline dataset more efficiently because a set data model is already in place. UGRC (the aggregator) and the county GIS offices (the stewards) adhere to the same standards, which streamlines the process.

Geospatial advisory groups exist to ensure that the wider geospatial community moves forward together. By taking an active role in one or more of these groups, GIS practitioners can ensure that their organizations are well represented and that the products they create fit into the larger geospatial fabric.

The geospatial fabric should stretch across the United States and around the globe. In other words, the authoritative data created at the local level should be the same data showing up in larger statewide, national, and even international datasets.

For example, comparing an address point in the National Address Database with its corresponding point in UGRC's statewide address database reveals they are the same. This is a successful steward-aggregator model in which local data makes its way to the state GIS office and then into the national fabric. This works because of collaboration, relationships, and shared data standards.

The goal of authoritative geospatial data providers is to weave open data together in such a way that individuals, organizations, and government entities have the tools they need to make better data-driven decisions.

A version of this story titled "Weaving the Geospatial Fabric of the World with Authoritative Data" by Greg Bunce originally appeared in the Spring 2022 issue of ArcNews.

PLANNING IS KEY TO BUILDING A COLLABORATIVE HUB: NC ONEMAP

Esri

NC ONEMAP HAS SERVED AS THE GEOSPATIAL DATA backbone of North Carolina since 2003. The initiative provides access to reliable statewide geospatial data that promotes public safety, informs government decisions concerning transportation and the environment, and increases economic vitality in North Carolina communities. It is an organized effort of numerous partners involving local, state, and federal government agencies, the private sector, and academia. NC OneMap is an initiative directed by the NC Geographic Information Coordinating Council (GICC) and maintained by the North Carolina Center for Geographic Information and Analysis (CGIA).

The NC OneMap parcel app provides parcel data from the 100 North Carolina counties and the Eastern Band of the Cherokee Indians into a statewide dataset with standardized attributes. The parcel data supports industries involved in, for example, economic development and emergency management.

CGIA recently chose ArcGIS Hub to build the next-generation NC OneMap portal. The team members invested time up front on content and site design and developed best practices for partners who share data. The team's planning pays off by giving users consistent results and an improved user experience and by generating fewer support requests.

NC OneMap promotes a vision for geospatial data standards and best practices; data currency, maintenance, accessibility, and documentation; and a statewide GIS inventory. It comprises 37 priority data themes, including statewide orthoimagery and aggregated parcel data from all 100 counties in North Carolina, plus lands of the Eastern Band of Cherokee Indians.

Since it was created, the portal has been updated to replace its legacy system and become a secure, collaborative space for the team's partners. CGIA needed a solution that would not only allow team members to quickly stand up their content but also be easy to maintain.

Customers of NC OneMap range from novices to experts. CGIA's challenge was to build a site that the average person would find intuitive. The portal now displays the most useful content at the top of every page. The portal also provides a reliable data search and discovery experience to customers. NC OneMap is an authoritative discovery point of statewide geospatial data supplied by different agencies. To ensure consistency, CGIA uses Hub, which allows organizations to capture and share data: create websites with stories, maps, and dashboards: and configure collaboration platforms in the cloud. With ArcGIS Hub, users can use existing layouts or design their own.

CGIA and many of its partners already use ArcGIS technology, and the close integration of Hub with ArcGIS Online was an important factor in the selection of Hub.

Several of CGIA's state partners, large and small, already have their own ArcGIS Online organizational accounts, so that prebuilt infrastructure allowed for quick integration. Because North Carolina state agencies such as the Department of Transportation and Department of Environmental Quality were already using ArcGIS Online in a production environment, the decision to use Hub for NC OneMap made sense.

Beyond technology, CGIA team members understood that planning was critical to their success. Since it's easy to begin building pages, it's vital to plan the site map layout before diving into development. Several years' experience with the previous version of the NC OneMap portal gave the team insight about customer needs and use patterns. The team approached the problem like a business, researching how people would use the site and get the most functionality out of it.

The team developed best practices for CGIA's partners to follow when sharing their data with NC OneMap.

The NC OneMap Guidelines, Recommendations, and Best Practices document was a key part of the team's solution. The guidance document improved NC OneMap usability, and it helped individual agencies and GIS professionals and users. "We know not everyone will come to the NC OneMap portal first for their data needs," said David Giordano, NC OneMap database administrator. "GIS analysts or professionals might already be in ArcGIS Online doing a key phrase search looking for data." The recommendations used for NC OneMap apply automatically everywhere users can access the data.

When partners provide a resource to the open data portal, one requirement is that they follow a standardized taxonomy system of tagging that resource. For example, open data will include the NC OneMap tag, tags associated with International Organization for Standardization (ISO) topic categories, and any other tags the data owner decides to include. The North Carolina GIS community already creates good metadata with ISO keywords, so following these accepted practices is relatively easy.

The site's organized content builds confidence and delivers a clean and consistent user experience. Partners follow best practices for sharing content, which improves the overall discoverability and usefulness of the information that individual departments provide.

Users such as foresters, engineers, land surveyors, and real estate agents access NC OneMap to look for data. CGIA responds to their feedback.

The updated portal receives fewer questions about how to search and find data than before. With NC OneMap 2.0, customers know what to expect and can easily find what they need.

A version of this story titled "Planning Is Key to Building a Collaborative Hub: NC OneMap" originally appeared on the Esri website in 2020.

CLIMATE CHANGE PROMPTS GRENADA TO CREATE NATIONAL DIGITAL TWIN

Esri

GRENADA, A NATION SMALL IN SIZE AND POPULATION, IN 2021 became one of the first countries to make a digital copy of itself—a 3D model that government officials can use for sustainability plans.

Like many island nations, Grenada confronts an uncertain future in the face of climate change. Increasing heat, intense rainfall, and saltwater intrusion into the water supply and soil have begun to threaten the country's two primary economies—agriculture and tourism. One challenge was how to continue to grow in a sustainable way and adapt to the changing environment. Addressing this challenge required a geographic approach—understanding what was happening and where.

Grenada's government had stores of raw geospatial data in the office of the Ministry of Agriculture and Lands. In 2019, the office received World Bank funds through the Regional Disaster Vulnerability Reduction Project and hired Fugro, a company that specializes in geographic and geologic data gathering and analysis, to do extensive aerial reconnaissance of Grenada. Fugro surveyed the Caribbean nation's three major islands, as well as six smaller ones. The result was a trove of information, including a lidar point cloud and extensive aerial images. But for practical purposes, there appeared to be no way to organize all this valuable information until the ministry used GIS technology to create a digital twin.

The power of two

The digital twin—a virtual representation of the objects and processes of a real-world system—has rapidly evolved in recent years.

The earliest digital twins were built to monitor the functioning of industrial factories, even to the level of individual valves and gaskets. Digital twins are now complex enough to model entire municipalities. City managers use them to monitor urban functions. Planners use them to visualize and analyze the effects of proposed changes.

Today's digital twins can be intricately detailed. Singapore's digital twin, for instance, extends to underground infrastructure and even includes some indoor features.

GIS experts within Grenada's government decided to extend the country's twin nationwide. Necessary for making digital twins fully operational, a GIS stores and displays disparate datasets that share locational components. These interactive and collaborative 3D models can then be used to drive better decision-making and policies at a larger scale than previous systems allowed.

A GIS enabled officials in Grenada to stack the imagery and point cloud data. The imagery and data could be consumed as separate map layers and combined to create something functionally larger than the sum of its parts. With its 20-centimeter resolution, the resultant aerial imagery produced a detailed representation of the island. Linking the 3D lidar data brought the imagery into full relief.

The digital twin goes deep

Government officials sought to use the country's digital twin to improve the lives of its residents, who are at the mercy of a swiftly changing ecosystem. The work started with extracting streets and buildings from the visual data so they could be sorted and quantified. The data could then be manually coded, but that process could require up to six months of work.

Grenadian officials worked with analysts from Esri to deploy AI capabilities within a GIS. A deep learning model was used to identify buildings. Within a day, analysts used the program to extract and label 55,000 built structures. They used geospatial artificial intelligence (GeoAI) capabilities to sort and classify other parts of the digital twin's visual data, such as roads, powerlines, streams, and other inland bodies of water, along with vegetation and land cover.

These classifications are valuable by themselves. For instance, staff from Grenada's Central Statistics Office, which partnered in the digital twin efforts, realized that the building data could simplify the process of planning the national census and recognized the value of having a complete building inventory of the country for the first time.

Combining the data categories created synergy. The Grenadian government and Esri used stream data, vegetation classifications, and digital terrain modeling (another segment of Fugro's aerial collections) to highlight spots in the country most in danger from landslides. This process was mainly automated: with classifications in place, the GIS generated the results. Other formulas and calculations produced flood susceptibility models, revealing where island residents were most vulnerable to extreme weather brought on by climate change.

The government of Grenada used imagery and point cloud data to model the potential impact of a 2-meter sea level rise in St. George Harbor, Grenada.

The 3D nature of the lidar data further contributed to the use of Grenada's new model. Seeing how far a building or road is from a landslide-prone area is helpful. Having the ability to zoom in and examine how a building is perched on a steep hillside or how a vulnerable road's angle of descent would appear from the perspective of a motorist, pedestrian, or cyclist adds context.

Seeing the future

The data from Grenada's digital twin is the basis for what the UN calls an integrated geospatial information framework (IGIF). It provides a complete view—realistic and integrated—of the country, which supports decision-making. This digital twin also has a predictive component that allows officials to visualize future challenges posed by climate change, along with solutions.

The government has used the digital twin and bathymetry information from Fugro to model scenarios for sea level rise—including storm surge and flooding damage—to see what will be impacted and where. The visual context of the map transcends numeric projections, facilitating policy making for prevention and mitigation.

The digital twin can also serve as an ongoing historical record. For example, the lidar data identified 4.5 million trees. If more aerial data is gathered at points in the future, the twin's GIS can analyze tree growth and note any significant deforestation. As important as AI capabilities are for this kind of calculation, they wouldn't have been possible without the lidar-enhanced imagery. Grenadian planners, interested in growing the country sustainably, can now look at a section of the map and imagine how further development will impact—and be impacted by—future changes in vegetation.

The value of a GIS-powered digital twin is that it enhances human observation. Although a digital twin can't literally see into the future, it is a window into several potential futures. However,

Grenada officials use ArcGIS Dashboards to conduct a population and housing census.

none of this would be possible without accurate imagery data and geospatial technology that ties geographic information together. These collaborative technologies are turning data into something meaningful, viewable, measurable, and actionable.

Working toward individual sustainability goals, other countries will likely follow Grenada's lead, building location-intelligent digital twins of their own.

A version of this story titled "Climate Change Prompts Grenada to Create First National Digital Twin" by Linda Peters originally appeared in the *Esri Blog* on June 16, 2022.

BP SHARES EIGHT LESSONS ON DIGITAL TRANSFORMATION

BP

L EADERS OF FORTUNE 500 COMPANIES BELIEVE DIGITAL transformation is the future of their business.

At BP, that transformation means embracing modern technologies, including digital, to improve its ability to monitor, predict, and optimize its business. This digital transformation will enable BP to create more sustainable operations, better manage expenses, and uncover new products and services for customers around the world.

Enacting digital transformation on a global scale is a vast undertaking at BP, a company of more than 70,000 employees dedicated to oil and gas exploration, production, transportation, refining, and retail—with airline and shipping divisions and a growing presence in biofuels and wind energy.

At BP, digital transformation can be viewed through the lens of location intelligence. Location intelligence helps BP professionals track the location and condition of assets, understand in real time the events that shape the regions and neighborhoods where BP operates, and identify areas of customer need and business growth.

Becoming a digital energy company required BP to overhaul its location intelligence capabilities. Central to that transition was modernizing its GIS capabilities. Lessons learned during that implementation can be valuable for any company that uses enterprise software to support digital transformation.

Lesson 1: Embrace the enterprise platform

Members of BP's exploration team use location intelligence, including GIS-based maps, to plot roads and design new access routes. BP's environmental teams use GIS to create impact assessments, and project teams use it to help design and deploy construction projects. GIS is the working geospatial brain that delivers the digital maps and analysis to help run a sustainable energy business.

As BP built a new foundation for location intelligence starting in 2016, the company systematically embraced an enterprise platform approach to GIS instead of building self-contained solutions for each business case.

Over the years, numerous instances of GIS had taken root independently at BP. Businesses had various levels of GIS deployments, and some GIS tools had been built for specific workflows. The data and workflows were disconnected because they were created by different teams at different times for different purposes, so BP had to support, maintain, and migrate software versions almost constantly, with a high cost of ownership.

Adopting a platform approach has helped reduce those costs. But cost was just the beginning of the benefits. For BP and other companies, an enterprise platform paid dividends in numerous ways:

- **The platform delivers faster time to value.** BP professionals want decision-support tools that adapt to their changing needs. With enterprise GIS, BP can more easily build mapping and analytics apps at the speed of business—in fact, BP enables business users to configure lightweight apps rather than enduring longer software development cycles from a central team. By letting the users create the tools the company needs, a business can quickly convert data and information into insights.

- **An enterprise platform speeds up collaboration.** When teams share an enterprise system, they can share the same location intelligence immediately, leading to quicker access, efficient analytics, and more informed decisions.

- **The platform creates a virtuous cycle for geospatial information.** A dataset that was considered basic information for one team can prove to be valuable to another team. For instance, activities that BP tracks in its upstream business are valuable intelligence to its downstream teams. These two parts of the business may otherwise have had limited interaction. But with shared data on a common GIS mapping platform, users can see the data resources available and use information from other teams. As a result, traditional organizational silos have started to dissolve. That, in turn, has led to new efficiencies, insights, and business opportunities.

On the enterprise platform, each business entity has its own space to manage data, workflows, analytics, and publishing. As a result, data and information are more accessible across the organization, leading to more opportunities to collaborate. The increased access to the data creates more value from each dataset and increases cycles of new capabilities and opportunities that may have been missed.

For these reasons and more, BP believes in providing a company-wide geospatial platform—rather than building specific solutions—that will lead to more sustainability and a larger return for an organization. The additional lessons explain some of the lessons and best practices BP has found while creating a successful and sustainable location intelligence system.

Lesson 2: When you're everywhere, you're nowhere

Location intelligence and GIS technology are used across BP, creating a paradox that companies often experience with other enterprise technologies: because the technology is needed everywhere, it is owned nowhere.

Historically in many companies, the IT department—because it serves all the company's businesses—oversees enterprise software platforms. Although this practice has been a successful model for enterprise tools with defined access and workflows, the model does not work as well for platforms with an open data structure and an almost unlimited set of use cases.

An enterprise platform, by definition, is a tool that can be used in many ways. In the case of a geospatial platform, the use cases are especially diverse. At various times and for various parts of the business, the platform can act as a system of record, a system of engagement, and a system of insight. BP believes a location intelligence platform should be placed in the business function as close to the core business as possible to benefit from business knowledge, workflow understanding, geospatial data skills, and competence in the toolkit.

Collaboration with BP's Information Technology and Systems (IT&S) team is still necessary for delivering the technology platform. IT&S helps deploy the right hardware, build the most efficient system design, manage network and storage needs, and safeguard the software portfolio.

In implementing its location-based platform, BP relied on a virtual geospatial project team composed of IT&S and business professionals to increase engagement, communication, and collaboration.

Deciding which part of the business organization should take ownership of the geospatial capabilities will depend on organizational structure, primary user groups, budgetary capabilities, and other concerns.

Lesson 3: Set your data free

Although the platform approach creates new opportunities by making data formats and services interoperable between company teams and functions, it also introduces new collaboration challenges.

BP has always had a mix of confidential data in its work. Historically, BP has locked access to much of its data, either because of limitations in the technology or a "better safe than sorry" approach. But this practice has resulted in duplication of data across projects and teams.

In the geospatial domain, this can cause problems. For example, BP found multiple files for the same pipeline locations, the same wells, and so on. The company also found subtle changes within some of the files. In many cases, the person who made the copy of the data moved on, and the team continued to use the original information, assuming it was the most current version.

At BP, about 95 percent of its geospatial data can be shared across the organization; only a few key datasets need to be locked. The solution was to challenge the status quo and open data by default, allowing it to be available and shareable for multiple teams and diverse workflows. BP advises other companies to secure processes and data where required but leave the rest open to allow exploration and innovation. An open system drives new workflows, insights, and the greatest value from the enterprise platform.

That said, BP also considered how to present data in the right way to help users of the open platform find the most appropriate information for their use cases. Simple summaries and basic data quality metrics help ensure that users are more informed when making data choices.

Lesson 4: Let business users extend the platform

Several years ago, Esri predicted that BP's users would soon create targeted apps for many of the company's location intelligence needs, and the prediction became reality. This evolution enabled almost anyone to build apps and workflows faster than the company could produce them in a central team, and this has made BP's business smarter and faster than ever.

Through today's digital software platforms, app development is lightweight, code-free, and user driven. These capabilities allow BP's community users to configure the platform to perform analytics and create insights at the pace of business.

This capability demanded another meaningful change in how BP deployed and supported GIS technology. BP opened the platform for its entire company. At rollout, the company gave business users a basic orientation on how to configure and create apps and then said, "Here's the playground. Do what you need to do to solve your business problems."

In response, BP sees hundreds of maps, apps, and dashboards emerging across its platform, many of which would only have been possible—under the prior way of working—after a user manually retrieved and reviewed various paper or digital projects and maps.

This capability to configure location intelligence apps has resulted in better decisions. Scaling that capability across tens of thousands of employees results in a diversity of ideas, an informed focus on integrated business workflows, gains in productivity, reduced risk, and business agility.

The One Map platform facilitates this collaboration daily. And yet there is a lesson lurking on the flip side of that revelation: once business users begin creating their own apps, the support team must keep up with them, and not vice versa.

Lesson 5: System and user support in this new paradigm

Typically, when a company deploys a software solution, the deployment team relies on an install manual and a standard database and format. The setup usually includes a support desk for users, with a few scripts to follow for common workflows. In the worst-case scenario, a vendor support desk might be needed to help with a software issue or tough workflow definition.

But with a geospatial platform, it includes various levels of technology, various platforms for deploying, a massive and diverse set of unknown workflows, and numerous data formats to oversee. Versions of information emerge over time, and each version can be valid, depending on the use case.

The team that supports an enterprise platform must be as knowledgeable as its user community. That means understanding the technology, the application use cases, data formats and workflows, analytics and displays, and even basic cartography.

Close collaboration between the IT and business organizations helps create the support structure needed to achieve enterprise platform success and sustainable results.

Lesson 6: Automate it

The success of BP projects tend to spawn similar efforts, which leads to more work for the team. This growth is a great indicator that the team adds value and the platform has a positive impact. But without good management, the team can consume its resources on ideas, leaving the core platform to suffer.

To counter that, the BP team created a rule early on: If there's a chance that a workflow will be needed again, automate it the first time. Automation helps on many levels. It ensures consistency and facilitates continuous improvements. It allows more transparent tracking and reporting of process status. Most importantly, it allows a company to run many more processes than could ever be achieved with people using keyboards. BP runs hundreds of automations around the company, with some active every few seconds and others only a few times a year.

The lesson BP learned was to invest as much thought in the automation platform as in the GIS technology itself.

Lesson 7: Marketing matters

When a company deploys an enterprise platform to tens of thousands of employees—whether it's GIS, enterprise resource planning (ERP), or another technology—branding, marketing, and communications matter. BP learned several marketing lessons during its digital transformation:

- **Build an identity:** To get traction on the big initiatives, create a program brand that users can identify with. BP called its GIS platform One Map, which highlighted the theme of a single mapping infrastructure for its location intelligence. The company then promoted the program identity, rather than the GIS technology behind it.

- **Mind your language:** At BP, as at any company, executives and professionals tend to think about GIS in the same way as PowerPoint or Microsoft Word. In other words, it's not the applications that are particularly important to them, but the business results that are. So, when BP began its enterprise platform rollout, it avoided the term GIS in promotional materials. In talking with the IT people, BP described systems of record, systems of insights, and platforms. In talking with the business sector, BP stressed production optimization, safety and efficiency, access to information, advanced analytics and dashboards, and overall quicker time to decisions.

- **Communicate:** The One Map launch campaign deployed emails, posters, and flyers in every BP office around the world, as well as advertisements on office televisions, to convey the business benefits of One Map and get BP professionals excited to use it. BP created three levels of internal websites. A website conveyed a generic message about location management and analytics and how those tools are used. Another site, titled Community of Practice, was tailored to BP's technical-savvy data and information professionals. Lastly, a tools dashboard allowed users to focus on tools, data models, and workflows in support of specific work. Communication is ongoing and includes regular Yammer posts, news stories, and internal articles. BP saw the need for a budget for communication, resources assigned to it, a plan for ongoing engagement, and an effort to end every communication with a call to action.

Lesson 8: Measure success

Measuring the impact of something as vast as a location intelligence platform for more than 70,000 employees spread across the world can be difficult, but it needed to ensure ongoing internal support. One simple way to gauge success is to monitor how many employees become active users of the platform and its apps. At BP, more than 7,000 users signed up to use the GIS platform before BP launched its communications campaign through social channels and other digital sharing.

Considering the diversity of use and the size of the global community, BP has taken its tracking and reporting to new levels of detail, including daily tracking of users, datasets, services, maps, apps, and more. These metrics help ensure the company can communicate the impact of the platform and support the system. With these details, BP can drive better hardware and software decisions, understand which datasets are used more or less than others, and see when a technical community is more active in one region than another, signaling that BP may need to organize targeted knowledge-share sessions.

Another way to measure success is by efficiencies gained. If someone at BP can access the data they need without having to load it manually, and if users can quickly configure the functionality needed to analyze and communicate it through maps, apps, or dashboards, the company might improve that person's annual efficiency by 3–5 percent. For a technical worker who needs location intelligence more often, it might be 5–8 percent. Since these gains come from a shared enterprise platform where an app or dashboard can be used in similar workflows, the increase is often greater.

When executives calculate productivity gains, factor the average resource salary and the number of people using the system, and follow progress over time, they can determine what a company gains from approaching location intelligence with a geospatial platform.

Digital transformation is not the destination but a way to solve business problems by using digital systems and integrated workflows. Considering the nature of digital systems, organizations must engage in an ever-evolving effort to update, maintain, and deliver needed capabilities.

BP's initial deployments were mostly internally hosted systems, but in envisioning the next generation of its platform architecture, BP has moved more infrastructure onto the cloud, so users can access GIS-driven data and apps on any device, anytime, anywhere.

As the technology has evolved, BP has noticed the industry's application of location intelligence change. AR and VR techniques are merged with BP's traditional location-based systems. The company has placed sensors on many assets, stationary and mobile, linking them to real-time IoT frameworks. BP's building information systems are also beginning to link with real-world geography.

Imagine the efficiency gains BP could achieve if an oil field technician could view real-time conditions, whether on a laptop, a tablet, or even an AR/VR display, in the office or in the field. In this real-time display, they can view equipment maintenance history, health status, operating parameters, user manuals, or even click a button to kick off a replacement activity—which integrates with the planning process and sends an equipment order to the procurement system.

Many executives might think, "It must have been easy for BP because they had millions of dollars and hundreds of people working on this location intelligence platform." In reality, BP had a small budget and only five people to start the platform. The team delivered to scale by rethinking some of BP's traditional approaches to geospatial capabilities. BP focused on platform over solution and worked to enable a larger community of users across the business who could drive BP's data, analytics, and apps forward.

For companies working to achieve sustainable digital transformation in the geospatial domain, the enterprise platform approach is worth exploration. BP found the approach to be an effective way to distribute technology throughout the company and multiply its benefits exponentially across its business.

A version of this story titled "BP Shares Eight Lessons on Digital Transformation" by Brian Boulmay originally appeared in *WhereNext* on November 12, 2018.

BUILDING A CADASTRE SYSTEM THAT MONITORS WORKING LANDS IN AZERBAIJAN

Esri

AZERBAIJAN, SITUATED IN A REGION THAT STRADDLES Europe and Asia, contains a striking range of terrain that moves through nine of the world's 11 major climate zones.

These distinct geographic variations make accurate mapping a challenge. But as Azerbaijan has continued its transition to a market economy, the need for accurate land valuation has intensified. The nation's government has approached this problem by embracing a merger of mapping tools and big data.

Using GIS technology, Azerbaijan embarked on a project to build a map that provides a clear picture of the land. Cartographers designed the map as a foundation for accruing more data—and more kinds of data. The data-driven map will help consistently refine the relationship between the land and its value.

The project, called the Land Cadastre and Registry System (TEKUIS), is managed by Layermark, a geospatial company based in Washington, DC. The effort to construct an accurate basemap of Azerbaijan included the step of depicting political divisions and a more rigorous effort to record individual parcel ownership.

"The first challenge was to develop the boundary management system, according to provincial, city, and local boundaries," said Yucel Tepekoy, Layermark CEO and TEKUIS project manager. "We started by modeling the land management system and implementing it with ArcGIS. This boundary management system includes land usage for governmental and private properties."

By 2020, the map depicted more than 8 million hectares (almost

20 million acres) of land. For maximum transparency, it remains open to the public. People can click any parcel on the map to see ownership history and create printable maps of any area.

As the basemap developed, Layermark turned to the next phase of the project, using GIS to describe soil conditions across the country. This data existed, but in some cases, it was as much as 50 years old. Soil conditions have changed since then, so the team needed to update its measurements.

To accommodate the incoming data, Tepekoy and his team designed a GIS as their information management system to keep the basemap up-to-date. "People can take soil samples from the field, send it to be processed, and then we use the data to generate soil maps," he said.

Of concern is the increase in soil salinity, which affects an estimated 40 percent of arable land in Azerbaijan. Water evaporating in the fields can leave behind water-soluble salts, making it difficult for some crops to grow.

The soil salinity problem is an infrastructure and modernization issue. When drainage systems fail, floodwaters can rise in the fields, resulting in increased salinity.

"Now, if there are parts of the land that have a salinization problem, the government can work with farmers to address the issue and decide where to grow crops to improve the soil," Tepekoy said.

The project involved generating good geobotanical maps, which detail pasture lands in Azerbaijan. The data on these maps helps farmers quantify the vagaries of animal husbandry. "They give decision-makers a better understanding of where the cattle should breed based on where protein values are likely to be higher," Tepekoy said.

The soil and geobotanical maps enable farmers to use their lands in more cost-effective ways. The maps also help the government more accurately calculate land value.

For a growing market economy, this kind of valuation is a

component of computing statistics such as GDP. It also strengthens the relationship between the government and residents. As the country weighs development projects that may require encroachment on private land, valuation maps help ensure fair compensation for landowners.

"If there are expropriation projects in a city that involve, for example, building a road or bridge on private lands, we can load that project onto the map," Tepekoy said. "And since we have all the cadastre data underneath it, we can show exactly which parcels are affected and help get the proper expropriation valuations to the owners."

Automating workflows

Although the map is the most visible manifestation of the TEKUIS project, the workflow systems that underlie it are just as important. What makes the map scalable is that the process can handle an enormous number of inputs.

"People were spending six to nine months to develop a fixed-area soil map or geobotanic map," Tepekoy said. "Now, they can use Esri maps with full automation. And it's a 100 percent web-based application. Without being shackled to a desktop system, data can come in from anywhere in the field."

Before the Azerbaijan mapping project was delivered, the work to create maps of soil administration and geobotanical data, as well as the boundary management system, involved a 98 percent manual process.

As the country continues to establish its role as a leader in European agriculture, Tepekoy said, "this project provides a quantum leap forward in terms of technology and automation."

A version of this story titled "Building a Cadastre System that Monitors Working Lands in Azerbaijan" by Brent Jones originally appeared in the *Esri Blog* on October 5, 2021.

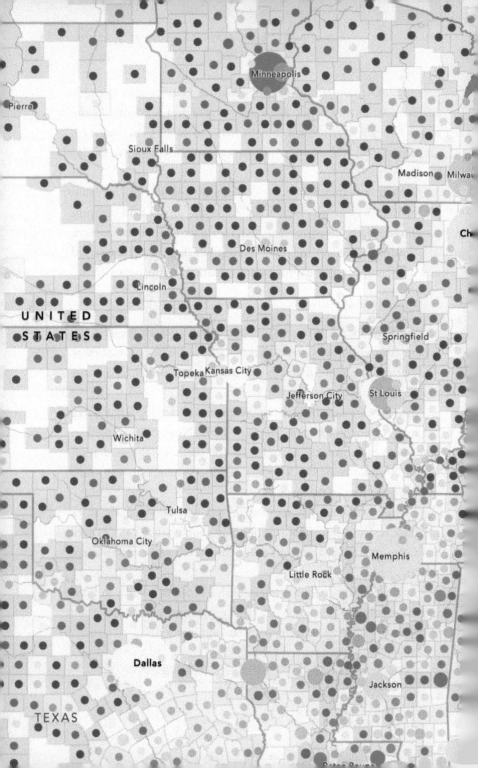

PART 3

ENGAGING
COMMUNITIES

O NE OF THE BEST WAYS TO ENCOURAGE A COLLABORATIVE
environment is to provide reliable and timely data and technology through frequent engagement that helps everyone learn about
and contribute to overall goals and objectives.

Geospatial collaboratives engage partners and their communities of practice through open GIS data, inclusive programs, and
policy initiatives. Portals and hubs serve as virtual destinations
that bring people together, providing practical tools and support
that facilitate teamwork, allowing users to interact and engage in
focused collaboration. Shared maps and apps, such as dashboards
and stories, inform those working to address today's issues and
challenges. Geospatial collaboration tools, such as surveys, map-based discussions, and open data portals, help users ranging from
strategic thinkers and specialized experts to local residents collectively add and review content, organize activities, and share their
stories.

For example, the nonprofit Western Association of Fish and
Wildlife Agencies (WAFWA) coordinates with agencies from 19 US
states and 5 Canadian provinces over 3.7 million square miles of
some of North America's most wild and scenic country to ensure
strategic, science-based conservation and practical resource management. WAFWA uses a collaborative GIS hub for open data
sharing and stakeholder engagement, making project information
more accessible to the public.

Geography and geospatial thinking are the best starting points for building a community engagement strategy, especially when working with agencies, stakeholders, and the public.

Community engagement, backed by a solid geospatial infrastructure, allows organizations to address significant local, regional, and global challenges that might otherwise seem too overwhelming or complex by presenting the information within a more understandable geographic context.

Today, organizations engage their communities to rally people around the issues and initiatives they care about most, such as

- responding to the impacts of climate change,
- creating sustainable business practices,
- protecting oceans, wetlands, and forests,
- finding equitable solutions for disadvantaged people,
- addressing disaster response and recovery,
- combating COVID-19 and other epidemics,
- rebuilding infrastructure,
- conserving water, forests, and biodiversity, and
- managing volunteer opportunities.

As a pillar for integrated geospatial infrastructure, community engagement supports collaboration, improves transparency, and builds trust.

Real-life stories

The real-life stories in this section explore how a variety of organizations engage their communities of interest to strengthen decision-making, improve services, and achieve better outcomes.

IN THE PHILIPPINES, A SHARED "DISASTER IMAGINATION" SUPPORTS RESILIENCE

Esri

S ITUATED AS ONE OF THE WORLD'S MOST DISASTER-PRONE countries, the Philippines routinely endures earthquakes, tsunamis, volcanic eruptions, landslides, and flooding. This archipelago of 7,640 islands in the Pacific Ocean resides along major tectonic plates and is at the center of a typhoon belt. Although its residents live in a region highly vulnerable to natural disasters, their government has set out to reduce the risk by mapping natural hazards and disasters. The mapping initiative will also support programs to thwart the impact of climate change.

"The Philippines is a good place to study natural disasters, but if you look at the human impacts, it's not so good," said Renato U. Solidum Jr., secretary of the government's Department of Science and Technology (DOST). Solidum also directs the Philippine Institute of Volcanology and Seismology (PHIVOLCS), created in 1951 in response to the catastrophic Mount Hibok-Hibok volcanic eruption. The event killed 500 people and spread devastation across more than seven square miles (19 square kilometers) of Camiguin Island. After the event, the government realized it needed to develop greater Filipino expertise.

Solidum and his colleagues at the institute lead a multiagency program, GeoRisk Philippines (GeoRiskPH), dedicated to mapping hazards and their impacts. They collect, analyze, and share risk-related data via interactive maps and apps using GIS technology.

"I told my team that we need to promote 'disaster imagination' about what the hazard can do," Solidum said. "Depending on our role in society, we have different perspectives on disasters. If we don't have the same imagination across the government and private sectors, we might be doing things that are not aligned, and instead of a collective effort, our efforts would go in different directions."

As the authoritative source for information about hazards in the country, the GeoRiskPH program provides a data clearinghouse with input from 21 national government agencies, 3 nonprofit organizations, and more than 50 local governments. The program provides apps and other tools to help safeguard the country's 108 million people from an array of risks.

For instance, among the first apps created by the institute was the PHIVOLCS FaultFinder, which shows where active faults are located; the government doesn't allow construction of houses or buildings on top of faults so the maps and information identify which land is unsafe for development. Interest in the app spurred a more ambitious project, HazardHunterPH, an application that enabled the Filipino people to see their exposure to any natural hazard, not just active fault lines, on a smart map.

"The real problem convincing people to undergo disaster preparedness activities is that they cannot imagine what can happen to them and their families," Solidum said. "We need to provide people with information not only of the hazards but, most importantly, what the hazard can do to them.

"There are more data now available across different organizations," he said, "and there is now a different state of willingness to share data. It was the right time to have a more integrative platform."

Decision-makers in the Philippines frequently turn to GeoRiskPH platforms for analysis to understand areas most vulnerable to ground tremors, tsunamis, severe wind, landslides, flooding, and volcanic eruptions. Using hazard and census information, GeoRiskPH

With the geodata system GeoMapperPH, smart mapping and spatial analytics solutions help users create apps and tools to support disaster risk mitigation, multiagency collaboration, and community engagement.

platforms allow for the analysis of risks to people based on location, age, and sex. Local governments can use this information to determine whether critical infrastructure is located in high-risk areas, create emergency management plans, draw safe evacuation routes, and decide where to site evacuation centers.

GeoRiskPH gives government leaders at all levels a view of the people exposed in a specific area. With data-driven maps, planners can determine which facilities—such as schools or hospitals—might be affected by a flood or tsunami. That awareness drives targeted risk mitigation and better disaster response preparation.

One component of the program's technology solution is Geo-MapperPH, the data collection mechanism that has supported information sharing among and across organizations, including several international partners. "They are able to see their data and input and understand that the tool will be more sustainable if they contribute," Solidum said.

The tool will also be used to develop the National Exposure Database, which will provide significant information for more detailed and accurate risk and impact assessments, and other initiatives.

The effort to collect hazard data is ongoing, with additional categories such as land use, economics, and demographics added. The institute is also working with other partners to add a social vulnerability index assessing the risks of disadvantaged communities and their climate-risk exposure.

"This platform is not simply an information and communications technology or geospatial platform, but, more importantly, it's a governance platform," Solidum said.

Visualization from GeoRiskPH maps helps streamline environmental compliance certification, giving developers quick access to the details and reports they need. The location intelligence guides insurance and investment decisions at large institutions.

Solidum hopes that the system will become more automated. Notably, researchers and government leaders in the Philippines plan to use this collection of data to prepare for impending climate events, especially sea level rise.

For this country, where disaster is already too familiar, climate change will exacerbate volatility. Solidum understands this reality clearly and is leading the move toward greater resilience.

"When most people talk about climate, it's mainly about weather, and they don't talk about the interaction of the ocean and the land, and what happens on the land," he said. "But that is where the impacts occur, and that's what we must prepare for."

A version of this story titled "In the Philippines, a Shared 'Disaster imagination' Supports Resilience" by James Miller originally appeared in the *Esri Blog* on November 11, 2021.

NEBRASKA'S ARCGIS HUB SITE BRIDGES GAP BETWEEN CITIZENS, GOVERNMENT

Esri

THE NEBRASKA OFFICE OF THE CHIEF INFORMATION Officer (OCIO) provides a range of technology services to state agencies, boards, and commissions. One of the OCIO team's objectives is to provide technology solutions that increase transparency and enhance collaboration. One of those solutions is the staff's development of NebraskaMAP, an ArcGIS Hub open data site that allows individual state agencies to share information with each other and the public on issues ranging from health and farming to transportation and the environment.

The searchable, user-friendly site provides access to shared GIS data and public web-mapping applications developed by state agencies. John Watermolen, former state GIS coordinator for the Nebraska Office of the CIO, said his goal while working for the state was to create "a sustainable, efficient, and functioning GIS ecosystem that provides a clearinghouse for quality Nebraska data." NebraskaMAP helped meet that goal, providing access for users from a central location.

The OCIO sought a technology solution to streamline the process of uploading information and keeping it current and to increase collaboration among state agencies.

The OCIO selected ArcGIS Hub in its recent version of NebraskaMAP. Watermolen and the GIS team were familiar with the product and knew that it had the features needed to integrate data and streamline operations. Hub provides a two-way engagement platform designed to connect governments and communities around policy initiatives to tackle pressing issues and share data, thereby increasing transparency.

The new primary geodatabase under OCIO is read-only and offers several views of data from state agency databases. Geodatabases provide a way to store and manage geographic information in ArcGIS Online, a cloud-based mapping and analysis solution. This data is then published and available for viewing on the open data site. Information is organized on the website in 18 different categories and has a simple toolbar for quick searching.

State agencies now operate under one umbrella using ArcGIS Enterprise, which helps users design maps and apps, manage data, conduct analysis, and share work. In the past, state agencies each had individual licenses for ArcGIS Enterprise and were working independently. But now ArcGIS Enterprise server licenses have been

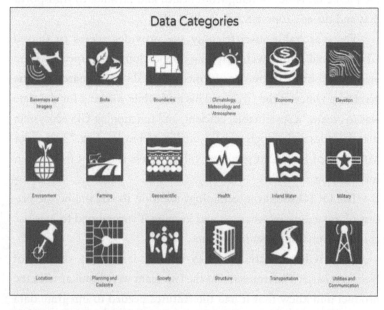

The NebraskaMAP portal, based on ArcGIS Hub, provides a two-way engagement platform designed to connect governments and communities around policy initiatives to tackle pressing issues, share data, and increase transparency.

consolidated and exist within the OCIO, facilitating the creation of the office's own ecosystem.

Stakeholders have seen a number of benefits since the implementation of Hub, with traffic to NebraskaMap averaging more than 200 hits per day. Hub enabled participating state agencies to contribute their own data, using their own geodatabases.

According to Watermolen, this capability gives agencies improved control over their data, allowing them to edit the data change schemas and determine privacy settings. Any changes are automatically pushed to the main enterprise geodatabase, ensuring that all published data is current.

He said that having data on one site eliminated duplicate requests to agencies. Data from multiple sources is accessible in one place, simplifying the user experience. For example, instead of asking several different agencies for information, users can now view data all at once.

"It's one place that the public can come to get accurate data and help [so that they can] be better informed, make better decisions, and have the best data available for any of their projects," said Watermolen.

The Nebraska Office of the CIO retired two virtual servers after the adoption of ArcGIS Hub. In addition, staff members no longer need to spend time updating data on the open data site because it's done automatically, allowing them to focus on other priority projects.

Additionally, Hub aided OCIO in its IT consolidation, putting all state agencies on the same platform and all servers in the state data center. Having one place for data increases the efficiency of operations and makes it easier to manage data.

"If different agencies participate in this manner, we can make this more of a true open data portal and not just a GIS-related one," Watermolen said. "We can build dashboards right into

NebraskaMAP to show how the data is being used. Our CIO likes the idea of the data coming alive."

A version of this story titled "Nebraska's ArcGIS Hub Site Bridges Gap between Citizens, Government" originally appeared on esri.com.

MARKING 50 YEARS, UNITED ARAB EMIRATES MAPS GROWTH AND QUALITY OF LIFE

Esri

T HE UNITED ARAB EMIRATES (UAE) HAS EXPERIENCED extraordinary urban and economic growth in the 50 years since it formed in 1971. The growth has been predominantly in Abu Dhabi and Dubai, where the bulk of the country's population lives. Abu Dhabi has seen the most dramatic urban growth of the seven emir-ates. The modern city features gleaming office towers, a multimodal transportation network, utility-scale renewable energy, and an edu-cation system that ranks in the top 20 in the world.

The UAE also has diversified its economy beyond oil, which required careful planning and discipline.

In advance of the 50th anniversary of the country's founding, the UAE's Federal Competitiveness and Statistics Center (FCSC) mapped the country and its people, using GIS to quantify how far it has come.

FCSC adopted GIS to modernize workflows and visualize the statistical indicators it gathers for such sectors as health, education, environment, and the economy. GIS data feeds 1Map, which isn't one map, but rather the mapping capacity for the nation, with a col-lection of feature layers such as roads, facilities, and demographics. The information allows residents to see how the UAE is faring and each ministry to see its strengths and weaknesses so investments can be made in the right places to improve quality of life for residents.

The discovery of oil in the 1950s fueled growth in the UAE, dis-placing pearls, fishing, and agriculture as key industries. Two decades ago, almost all the country's economy was oil based. Now oil makes

As with many measures of UAE's modernization, the diversification of the country's economy away from oil dominance is carefully tracked. Dashboard screenshot courtesy of FCSC.

up less than 30 percent of GDP. A series of plans have guided national investments away from oil dependency.

His Highness Sheikh Mohamed bin Zayed Al Nahyan, the third president of UAE, spoke in 2015 about this effort and the steady focus toward a future without oil. "In 50 years, when we might have the last barrel of oil, the question is, when it is shipped abroad, will we be sad?" he asked. "If we are investing today in the right sectors, I can tell you we will celebrate."

UAE has diversified with investments in infrastructure, hospitality and tourism, and technology. Three sovereign wealth funds in UAE invest on behalf of the government, including Abu Dhabi Investment Authority, with assets of nearly $700 billion. This strategy of investment outside of oil aims to provide long-term returns from this nonrenewable resource.

Many of the investments are tied to revenue diversification and improvements in quality of life, including the goal of being a

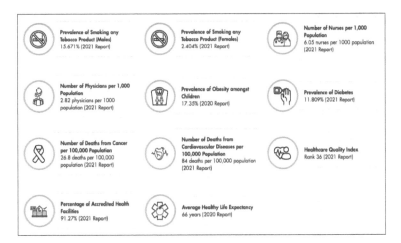

These data categories represent some of the many indicators compiled to assess the health of UAE's population. Many of these indicators were also mapped. Screenshot courtesy of FCSC.

destination for world-class health care, which illustrates how statistics drive societal advancements. First, the UAE needed to know the health of its people, and then it needed to compile details on healthcare facilities.

"We started the UAE Facilities Catalog by collecting the health facilities datasets along with the indicators such as how many are publicly owned, how many beds they have, and the number of medical staff and physicians," said Marwa ElKabbany, a GIS expert for FCSC. "We mapped it to feed our geostatistical platform, then we overlaid facilities with our population and administrative maps to understand and evaluate their geographic distribution."

FCSC has been mandated by the Ministry of Cabinet Affairs to collect data from federal government entities in a seamless manner, including location data where applicable while fostering best practices regarding data quality and standards. The 1Map name and branding by FCSC matches the country's data integration goal, which fits the capabilities of modern GIS to tackle challenges across sectors.

The FCSC GIS team works with federal and local partners to implement a national collaborative geographic information portal that collects geospatial data from partners, processing the data to harmonize and standardize it, and then creating different applications and sites to serve the community and governmental partners through the one-stop 1Map portal.

The creation of the National Digital Data Ecosystem is a key priority of the UAE government. FCSC has been working on a variety of projects as the lead federal entity for this effort, such as the Data Maturity Index, designed to enable agencies to manage and administer their data in accordance with international standards and best practices to ensure accessibility and data flow, governance, and quality. Another example is the Emirates Data Network project, which supports the exchange of federal administrative data contributing to a national knowledge hub.

At FCSC, ArcGIS Enterprise enables integration across each ministry's implementation of ArcGIS software through portal-to-portal collaboration. 1Map provides the foundational maps that each ministry builds on with its own data that it can share through ArcGIS Enterprise while maintaining authority and security over the data it creates.

1Map datasets include population distribution; land use and housing; public facilities such as hospitals, schools, and cultural and religious destinations; and preserved natural areas and parks. This data is further broken down into components. For example, population data includes demographic details such as gender, citizenship, and age at different scales. Maps show each of these indicators as well as population density across the country.

Much of the 1Map data is sensitive and restricted. Although

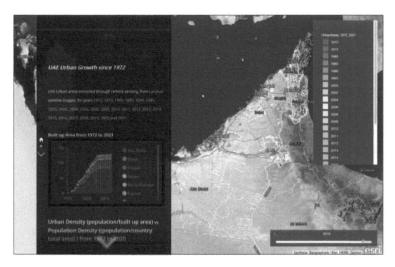

This map depicts the growth of urban areas since the country's founding in 1971.

some datasets are public, others are controlled with access given only to certain ministries or approved researchers.

The data is displayed in dashboards built with ArcGIS Dashboards to show progress on issues that are specific to a ministry's objectives or that are relevant for the nation. For instance, a public-facing dashboard gathers total international nonoil trade for the country. Narrative maps have been built with ArcGIS StoryMaps to better communicate with the public, such as a story that uses satellite images of Earth at night to illustrate urban growth in the country. Data is organized using Hub to indicate progress, including the 50 Years of Prosperity project. Specialists further analyze data to inform leaders and check on the progress of policies—and to craft new ones—all focused on moving the country forward.

Competitiveness on quality-of-life issues is central to the way FCSC operates. For instance, an initial calculation by international organizations about rural access to roads showed that three-fourths of populations living in rural areas had access to an all-season road within two kilometers.

"The roads in UAE are amazing, so it didn't make sense," ElKabbany said, relating how the World Economic Forum report in 2019 ranked UAE seventh in the world for quality of road infrastructure and the Legatum Institute ranked UAE first for user satisfaction of roads and highways in its Prosperity Index Report.

"We looked at the dataset the international organization used to understand it, then we consolidated rural and urban boundaries from local municipalities, built a national population grid reflecting local statistics for population estimates, and overlaid all this data with the roads network according to the World Bank methodology," ElKabbany said. Within three years, 99.54 percent of rural-area residents had access to a road within two kilometers, she said.

The map helped verify the accuracy of the data and how the data helps verify the accuracy of the map. The road access indicator showed how rural and urban area delineation wasn't accurate and how population estimates were off. The climate and geographic conditions in the country favor urban development where housing is weather controlled and there's easy access to health care, schools, and other services, according to ElKabbany.

Official statistics for the UAE, as with many nations, have long been maintained in spreadsheets, but there's a growing move to put these measurements on a map, thanks to the UN Sustainable Development Goals (SDGs) adopted in 2017. The SDGs recognize intertwining issues, such as how alleviating poverty goes hand in hand with improving health and education and spurring economic growth. Putting indicators on the map provides a crucial view to address inequalities and pinpoint where to limit the impacts of climate change.

When the COVID-19 pandemic hit, FCSC had just launched

maps of health facilities and population distribution. That data identified areas of risk and helped leaders understand the importance of maps and geostatistics for crisis management.

FCSC also uses satellite imagery and machine learning to fill in data gaps. Its team has collected building footprints across the country and used smart meter data about electricity and gas consumption to derive population estimates for where people live and work. FCSC continuously monitors nontraditional data sources such as anonymized phone location data to understand people's patterns of movement. Analysis of movement patterns helps compare facilities and opportunities across different geographic areas, which supports planning processes.

"We have to be ready. We must prepare our data, and we need to measure prosperity and competitiveness to ensure high quality-of-life standards are maintained and elevated," ElKabbany said.

A version of this story titled "Marking 50 Years, United Arab Emirates Maps Growth and Quality of Life" by Linda Peters originally appeared in the *Esri Blog* on September 20, 2022.

ARCGIS HUB ENABLES COMMUNITIES TO RAPIDLY SHARE UP-TO-DATE DATA ON COVID-19

Esri

TO INFORM THE PUBLIC QUICKLY ABOUT SWIFTLY changing information throughout the coronavirus disease 2019 (COVID-19) pandemic, scores of governments and agencies used ArcGIS Hub to create open data sites. These sites helped increase transparency and built trust in local government—and they were important during a large event like a pandemic to communicate up-to-the-minute news and resources. With maps that show everything from community spread of the disease to updates on local school closures, these open data sites provided residents and local officials with the information they needed to protect their communities, stay safe, prevent infections, and make data-informed decisions.

This story includes several examples of open data sites, built with ArcGIS Hub, from across the United States that have given residents important resources and updates on COVID-19 throughout the crisis.

State of Maryland

To coordinate content and updates to the community, the Maryland Department of Information Technology (DoIT) Geographic Information Office (GIO) created a site using ArcGIS Hub in partnership with several agencies, including the Maryland governor's office, the Maryland Department of Health (MDH), and the Maryland Emergency Management Agency (MEMA). The hub site was designed to be an authoritative source of reliable information on COVID-19 and provide users with fast and easy access to resources.

"I want to instill confidence in Marylanders that their state government is providing accurate and timely data to all visitors to the site," said Julia Fischer, former director of data services for business intelligence and GIS at DoIT. "The data is accessible to all Marylanders and is providing a clear and definitive picture of the state's response to the unprecedented events occurring right now."

Fischer said the DoIT GIO team selected Hub to create the new site because they were familiar with the product. The team members created the initial site in about four hours, coordinating with staff at MDH, MEMA, and the governor's office so they could make content updates directly to the site.

The site included a prominently placed dashboard that shared vital statistics regarding confirmed COVID-19 cases (broken down by county and zip code), hospitalizations, and testing results. The site also provided details regarding symptoms of COVID-19, best practices for social distancing, links to local health departments, and more. It featured practical resources such as links to information from the Centers for Disease Control and Prevention (CDC) and related highlights from governor's official website.

"We are providing a site that includes quick access to interactive maps and dashboards, as well as data analytic capabilities, ensuring that Marylanders are well-informed about health and safety topics related to COVID-19 in Maryland," Fischer said during the pandemic.

MDH consolidated data to gain a clear picture of the statewide efforts to combat COVID-19, along with their effects. As such, the data featured on the site was derived from the work of MDH and its partners. Additional subpages were developed to show the locations of congregate facilities and testing sites throughout the state to help visitors understand the impact of the pandemic.

"Transparency of information is everything," Fischer said. "Governments have an obligation to communicate data and empower all of us to make informed decisions for ourselves, our families, and our communities."

Montgomery County, Pennsylvania

Officials in Montgomery County, Pennsylvania, established a site with ArcGIS Hub when the county recorded its first case of COVID-19 in early March 2020. The initiative began with a simple storytelling map that showed the location of COVID-19 cases in the state and the county and then evolved into a full hub site with maps, resources, and information on the pandemic.

David Long, county GIS manager, said Hub integrated seamlessly with the capabilities of the GIS division's existing ArcGIS technology.

The Montgomery County hub site provided basic but vital information about COVID-19, including prevention advice and symptoms of the disease. The site also displayed county-specific statistics such as the age, gender, and municipalities associated with COVID-19 cases, local county news, and updates from Governor Tom Wolf, including his plan for reopening the state. In addition, the hub hosted public testing registration forms for residents to preregister for a COVID-19 test.

A range of resources were available, including information from the CDC, links to the Pennsylvania COVID-19 site and the Pennsylvania Department of Health, and the Johns Hopkins University dashboard that showed global cases.

A stand-alone business page, accessible from the site, was created for the Montgomery County business community. This page included information, such as how to apply for business loans, and a capability that enabled businesses to indicate whether they were open or closed or offering discounts. With help from Google Translate, the GIS team made the main page of the hub site available in Korean to serve the county's large Korean-speaking community.

For Long, an additional benefit of Hub is that it responds well to the mobile environment.

"ArcGIS Hub has made it simple to continually update our content," he said.

Delaware County, Indiana

In Delaware County, Indiana, the Delaware County GIS Department managed a COVID-19 hub site in partnership with the Delaware County Emergency Management Agency and the Delaware County Health Department.

The local coronavirus hub site served as the official one-stop source for the community's response to the COVID-19 epidemic, said Kyle Johnson, Delaware County's GIS department coordinator. Johnson had used Hub on a different project and found it integrated well with ArcGIS Online content, so integrating Hub into the county's COVID-19 site was simple because of its drag-and-drop interface.

"Providing fast, important information to the public is always a challenge, especially during emergency situations," he said, adding that Hub and ArcGIS Online made the job of disseminating information to the public much easier.

The county's Emergency Management Agency, which tracked self-reported illnesses and recoveries, was the primary data source for the site. A dashboard also gave users access to the State of Indiana's dashboard and the Johns Hopkins University global dashboard.

When food distribution programs began in the county, the GIS department also included maps on the hub that showed the locations of the distribution centers. Additional pages on the site focused on local resources, including details on school closures and student food pickup sites and daily briefing videos from the county's Emergency Management Agency.

Having the COVID-19 information in one place kept the

community informed on the pandemic, and the hub received a lot of positive feedback, he said.

University of South Florida

The Digital Heritage and Humanities Collections GIS team at the University of South Florida (USF) in Tampa, Florida, developed the Florida COVID-19 Hub site with a regional and statewide focus. The site was designed to provide a platform for data access, research, and visualization.

The project helped serve campus researchers and engaged a community of learners and educators, as well as the broader local and regional areas, said Lori Collins, research director and principal investigator.

The GIS team chose Hub because of its ability to host multiple types of apps and data. Also, a number of agencies and data creators in Florida were already building GIS dashboards that the team knew could be quickly incorporated and expanded using a hub approach.

"We chose to not duplicate efforts but be inclusive and collaborative from the start, making ArcGIS Hub an obvious choice for building out an application like this," said Benjamin Mittler, GIS manager and lead developer and designer. "ArcGIS Hub allowed us to easily harvest data from other sources and display [it] in a simple way. Additionally, the Hub interface allowed us to add several important elements to our page with little to no coding."

Collins added, "We wanted to provide people with a tool where they could just sort of easily land and get the resources they need to make on-the-ground decisions but also predictively look at things. ArcGIS Hub provided us with an easy mechanism to serve data out that way and to continue to add and grow the platform as new needs arose."

The Florida COVID-19 Hub site included the hub app, four

custom dashboards, one web app, nine web maps, two feature services, and nine feature services streamed from government agencies. The statistics, map, social media, and featured applications sections of the dashboard were all created using built-in Hub functionality. The site also included a map that displayed the Florida case count by county, with symbology that emphasized the largest outbreaks. In addition, a custom layer examined cases and deaths related specifically to nursing homes.

Resources on the site included embedded Twitter feeds for the Florida Department of Health, USF, Hillsborough County, and Pinellas County to ensure that users received real-time information on local and statewide closures. Cards on the site linked to local and national news sources for additional information.

The hub also connected to two sources of spatial data. The first was an ArcGIS Online group site with curated and updated data relevant to COVID-19. The second was a storage folder that the Department of Health used to export and archive data downloads each day.

USF libraries, which the GIS team is part of, plan to help with metadata and paradata creation. This work will allow people to reuse the data in the long-term and conduct continued studies of the impact of the pandemic in Florida.

"Our ArcGIS site has become a really good way to collaborate and be a part of the community that we serve," Mittler said.

A version of this story titled "ArcGIS Hub Enables Communities to Rapidly Share Up-to-Date Data on COVID-19" originally appeared in the Summer 2020 issue of *ArcNews*.

PRAGUE: EXTREME-HEAT EVENTS SPUR CLIMATE ACTION, USING GEOSPATIAL TECH

Esri

THE *PRAGUE DAILY MONITOR* FEATURED THIS HEADLINE IN early July 2015: "Extreme Heat Wave Has Hit Czech Republic." The story said currents of hot air moving from Africa were driving up temperatures across Europe. It was the first of four heat waves in the Czech Republic in 2015. More than half of the days in July and August that year recorded extreme temperatures, breaking a record set in the country during a similar string of heat waves in 1994.

Two years after the 2015 heat waves, the City of Prague, the country's capital, issued a document that outlined a four-year plan, beginning in 2020, to "enhance long-term resilience and reduce vulnerability to climate change." To meet these objectives, officials from the Prague Institute of Planning and Development (IPR Prague) adopted a big data approach. They make extensive use of GIS, which allows them to understand how Prague is reacting to the climate crisis in the present while also devising ways to meet the climate-related challenges of the 21st century and beyond. GIS allows IPR Prague to view and analyze the city in both its street-level granularity and its bird's-eye view totality.

Prague is vulnerable to extreme heat. Compared with other European cities, it has more paved spaces, built-up areas, and industrial infrastructure—the kind of spaces that can create so-called heat islands. But Prague also contains a lot of green space and vegetation, areas that offer respite from the heat. From a planning perspective, this tapestry of extremes presents a puzzle to be solved so that Prague residents are equipped to deal with global warming.

Coastal communities get a lot of attention because of the

challenges they face in addressing sea level rise related to climate change. The reality is that large cities—even those that are land-locked—also face big challenges as they adapt to climate change.

The population density and development in urban areas exacerbates the effects of rising temperatures. The economic and social diversity of cities means that certain communities feel the effects more than others. The sheer enormity of a city means these effects are themselves diverse, varying widely throughout the city.

For these reasons, cities are increasingly taking proactive approaches to climate change that supersede national policies. In 2018, the Czech government declared that climate change mitigation would be a national priority, a year after Prague released its strategy document.

"For the Czech government, it's always seemed like more theory than practice," said Jiří Čtyroký, director of spatial information at IPR Prague.

Key to this strategy is a commitment to use data in a way that helps IPR Prague understand how climate change affects Prague, how these effects will evolve over time, and how to best develop the city to meet these challenges.

"What Prague is doing fulfills the criteria of the national government but goes much further," Čtyroký said. "We have an implementation strategy with really ambitious goals, including making Prague completely carbon neutral by 2050."

Sensors throughout the city measure variables such as temperature fluctuation, solar radiation, and humidity. "There are more droughts, less precipitation, and more tropical temperatures than ever before," Čtyroký said. "It makes the city streets and public spaces less livable and more stressful for people." But which streets and spaces? Which people?

IPR Prague integrates information from the environmental

sensors with health and demographic data. For instance, IPR staff can see heavy concentrations of young children and the elderly—two populations at increased risk from hot temperatures.

A study that compared the heat waves of 1994 and 2015 found that during the summer of 1994, mortality rates among the elderly increased at about the same rate as younger populations—yet, in 2015, they were significantly higher. The theory behind this shift is that positive socioeconomic changes since 1989's Velvet Revolution, when nonviolent student-led protests led to the fall of the communist government, had made those who were age 64 years and under "less vulnerable to heat stress over time."

But during the same 21-year period, the country's over-65 population increased. They were, as a group, still vulnerable to heat waves, and their greater number counteracted the gains made by younger groups. On balance, they were the reason the total mortality impact of the 2015 heat waves was greater than it had been in 1994.

GIS provides a way to visualize—and therefore contextualize— these statistics. Demographics and other human population data become layers on a smart map. The layers can be set against environmental features of the city, offering graphic representation of how the city and its populations interact.

This effort grew out of requests from Prague officials for IPR Prague to devise a way to rate the viability of future projects. "We were asked to develop an aggregated map that showed the best areas to expend effort and money," Čtyroký said. "That was the beginning of our vulnerability index."

This index, created by bringing together the various data sources through GIS, helps identify at-risk areas. Čtyroký highlighted Old Town—Prague's original city center, which dates to the medieval walled city—on the map. "It's one of the worst spots in terms of climate," he said. "It's a densely built-up quarter, and there's no

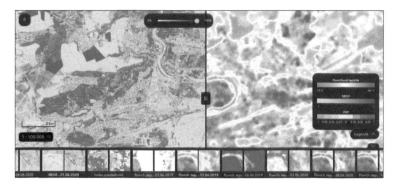

A tool to view surface temperature and vegetation density side by side helps determine the need for shade.

wind—just the cobblestone, stone, and asphalt surfaces." This area could benefit from increased green spaces and tree cover.

"The plan is to update the index every few years to see how the situation has changed," Čtyroký said. "It won't be a static thing— and hopefully it will be improving."

Prague's climate strategy also involves using GIS to construct intricate 3D models of the city's microclimates. Once established, these models will provide a way to analyze the effects of mitigation strategies, before the city makes any large investment of time or money.

"We'll use them to improve our proposals," Čtyroký said. "For something small, like replanting trees, we won't bother to model it. But if we've got a huge redevelopment project, we'll want to model it right up to the final stage of the proposal." These models also provide a way to communicate plans with other government agencies and the public.

"This is an interesting area," Čtyroký said, pointing at the map. "It's basically a 19th-century development, an old industrial and residential quarter. And you can see it's a relatively climatically okay region so far—mostly brick and grass."

The interactive 3D view can be colored and explored for different purposes. Here, the colors relate to building heights.

Čtyroký zoomed in for greater detail. "But what's this?" he asked, pointing at a spot. "It's one of the biggest brownfields in Prague—an old railway station and warehouse, and they'll be completely redeveloped into a new quarter for 30,000 people. And that's exactly the kind of place where we'd want to implement the microclimate modeling, because it's a place where massive change will occur, and we want to see how it will work."

Prague's climate strategy also involves adding even more data sources, working in conjunction with the country's ministry of the environment. "We'll join our databases related to environmental indicators and manage it as one big project," Čtyroký said.

As the data flow increases, the GIS-enabled map of Prague will become more complex with a multiplicity of applications. Looking toward Prague's carbon-neutral future, city officials are discussing with IPR Prague the possibility of rooftop photovoltaic and wind-power generation.

IPR Prague is also exploring becoming more involved with a project under way to study energy consumption in municipal buildings. In effect, this would mean expanding the map to include and model indoor spaces.

"We'd like to figure out how to combine information that we know about surface temperatures, weather, and heat waves and discover how these relate to energy consumption in the buildings," Čtyroký said.

A version of this story titled "Prague: Extreme-Heat Events Spur Climate Action, Using Geospatial Tech" by Richard Budden originally appeared in the *Esri Blog* on April 27, 2021.

PORTUGAL'S SECURITY SERVICES SHARE DATA, CONFIGURABLE APPS TO AID PUBLIC SAFETY

Esri

THE SEPTEMBER 11, 2001, TERRORIST ATTACKS IN THE UNITED States were one of the reasons that national security became more important to governments around the world. Many governments realized the importance of creating a collaborative environment that security organizations and agencies could use to ensure the safety of their populations.

Portugal's Ministry of Internal Administration, known by its Portuguese acronym MAI, is responsible for public security and emergency management and supports the electoral administration, road safety agency, and immigration and refugee affairs. With an increased focus on national security, MAI was compelled to get Portugal's security forces and services to work in a more coordinated and integrated manner.

To that end, Esri Portugal, Sistemas e Informação Geográfica, S.A., and telecommunications company Altice, in collaboration with MAI, built a geospatial platform called GeoMAI that allows security personnel from a range of organizations to integrate data to identify the locations of dangerous situations, understand how they are unfolding, and learn how to respond quickly to stop them. The platform aggregates ArcGIS Enterprise, ArcGIS Online, ArcGIS Pro, and several apps—along with solution engineering and consulting and training services—to make gathering, combining, and visualizing disparate data relatively swift and simple.

GeoMAI's main objective is to make data from multiple sources and systems (internal and external) available to MAI and have various solutions to help staff analyze and act on that data. The system supports the country's planning, prevention, and operational security and safety services, including the National Republican Guard, Public Security Police, National Authority for Civil Protection, National Road Safety Authority, and Immigration and Borders Service.

But GeoMAI is more than an information platform. It makes available tools and solutions that increase the efficiency and effectiveness of data integration when it comes to analyzing risk and understanding social criminal phenomena. It is also a tactical support tool that ministry staff, police, and the general public can use to combat crime and create public security policies.

When the project began in 2014, Esri Portugal's sector lead manager for defense and security thought that, because of MAI's specific needs and requirements, GeoMAI would need a strong app development component. But as ArcGIS technology evolved, improving templates and configuration tools, the project focus changed to have GeoMAI rely more on configured information products. This capability would significantly increase the speed at which data and apps could be delivered to meet the demands of the ministry and its public safety and security agencies.

One challenge was that Portugal's security forces and services were at different maturity levels in their ArcGIS capabilities. To remedy this, the team selected various ArcGIS software components—including web apps and dashboards—to adapt to each organization's needs. For instance, Esri Portugal worked with MAI to create geoprocessing services published in ArcGIS Enterprise to report all sorts of information about various incidents, such as which species were harmed and which economic sectors got damaged in a wildfire burn area. With this data, for instance, the National Republican Guard can generate maps, map stories, and printed materials to give to other entities for environmental evaluation. These organizations can then use visuals to highlight environmental statistics and develop wildfire prevention measures.

MAI and security forces also share a secure network called the National Homeland Security Information Network, known by its Portuguese acronym RNSI, that provides network and telecommunications services and map services with basic geographic information for Portugal. Ministry officials and security forces can use the same orthoimagery, administrative boundaries, and street networks as the basis for their maps.

As the project progressed, Esri Portugal collaborated with the agencies, in coordination with the ministry, to analyze their capabilities, needs, and potential.

With the National Authority for Civil Protection, for instance, Esri Portugal started by building up its operational capacity. The team then implemented ArcGIS Dashboards so the agency could monitor incidents and assets in the field, instituted ArcGIS Web AppBuilder so staff could create web apps that support team coordination, and set up ArcGIS Collector and ArcGIS Survey123 to help staff gather information in the field. Once these components were in place, the National Authority for Civil Protection could configure its own solutions, using ArcGIS Pro to harvest, sort, and manage incoming data.

For the Public Security Police, Esri Portugal set the organization up with Dashboards and ArcGIS Enterprise Sites. Although neither of these is fully functional yet, the police will be able to use them in conjunction with Web AppBuilder to create tailored web apps and pages that show criminal statistics and various kinds of business information.

For the National Road Safety Authority, Esri Portugal began by using ArcGIS API for Python to develop algorithms that could standardize and validate information gathered about accidents. The team then made it possible to present this data using Dashboards. With these capabilities, the agency can monitor traffic incidents on a dashboard, making it easier to manage traffic and road safety.

Through GeoMAI, Portugal's internal administration ministry and the safety and security agencies under its auspices can more easily and comprehensively visualize data, which helps them contextualize security challenges and obtain better results.

A version of this story titled "Portugal's Security Services Share Data, Configurable Apps to Aid Public Safety" originally appeared in the 2019 Summer issue of *ArcNews*.

Hermosillo

PART 4

BUILDING CAPACITY

COLLABORATIVE GEOSPATIAL ORGANIZATIONS integrate spatial data, technologies, supporting systems, and processes to enable informed decision-making across sectors, academia, nonprofits, and levels of government. A foundational component of integrated geospatial infrastructure is capacity building, which grows a knowledgeable geospatial workforce and community of users skilled in the use of GIS and data.

Capacity building involves training, education, mentoring, professional development, and other types of support that help individuals acquire necessary skills and knowledge. Organizations strengthen their communities by expanding the capacity of people to contribute and respond to complex problems using geographic information. Organizations train staff, stakeholders, students, and volunteers and provide affordable access to geospatial data and tools. They nurture developers and business talent and mentor the next generation of leaders and users so the value and benefits of the geospatial ecosystem are sustained for future generations.

By building capacity, organizations ensure that maps, geospatial analysis, and geospatial thinking become a regular part of decision-making and finding innovative solutions. Community members equipped with geospatial skills and tools can effectively use geospatial data and technology to build knowledge and understanding and formulate action.

For example, the American Red Cross focused its service delivery using GIS tools to provide a common operating picture across its many partners and participating agencies. As its capabilities progressed, the Red Cross increased capacity within its organization and collaboration with its operational partners. Using geospatial infrastructure helped the Red Cross more easily assimilate open and secure geospatial data into its daily operations, leading to better assistance to its clients and creating more resilient communities.

At a deeper level, capacity building helps establish best practices for security, privacy, accessibility, open data standards, and data quality, which, in turn, provides a smooth and diverse content flow through the geospatial infrastructure.

Organizations build capacity to support their communities in several ways, including training and education, resource sharing, outreach and communication, collaboration and partnerships, and volunteerism. Today, organizations build geospatial capacity to

- grow digital economies and innovation,

- engage scientists and communities through data sharing,

- use data and technology to understand challenges and contribute to solutions,

- implement geospatial strategies and initiatives, such as national spatial data infrastructures (NSDI), geospatial ecosystems, and open data strategies,

- facilitate seamless data development, information sharing, and collaborative decision-making across sectors of the economy,

- engage with partners to achieve the UN Sustainable Development Goals (SDIs), and

- sustain the value and benefits of geospatial investments.

Capacity building helps organizations fully realize the benefits of their geospatial infrastructure.

Real-life stories

The real-life stories in this concluding section explore how capacity building supports communities of practice to connect content providers and create a broader base of geospatial content consumers and users, including decision-makers, researchers, students, developers, and an engaged public.

WHITE HOUSE PORTAL HELPS COMMUNITIES ASSESS EXPOSURE TO CLIMATE HAZARDS

Esri

FINDING COMMON GROUND ON MOST COMPLEX ISSUES— including how to collectively act on our changing climate—starts with a foundation of shared knowledge, expertise, and information.

As the United States advances on its climate action goals, the White House is using a tool—developed with the National Oceanic and Atmospheric Administration, the Department of Interior, and the US Global Change Research Program—that offers scientific data to help us better understand the impacts of climate change at the local level.

The Climate Mapping for Resilience and Adaptation (CMRA) Portal helps cities, counties, states, and tribes make better decisions about where and how they need to act. Central to the portal is the CMRA Assessment Tool, which helps users explore current and projected climate conditions where they live and work.

After the 2021 passage of federal legislation to overhaul infrastructure in the United States, the country aims to make sure new and renovated roads, bridges, railroads, power grids, water systems, transit routes, airports, and ports achieve climate resilience in all communities.

The CMRA Portal builds on years of scientific knowledge and investments in geospatial platforms, but in the past these resources have been difficult to find and understand because they were not integrated to provide a complete picture. Using GIS with the portal addresses that challenge, integrating information in the context of location.

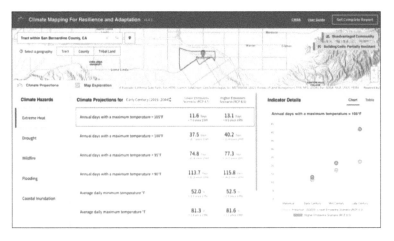

The portal gives users a national view of climate projections for the United States, including an analysis of extreme heat potential in specific areas.

The portal brings the data together into one place. Environmental data and the social and economic factors affected by climate-related hazards can be explored in the form of maps, charts, and reports by anyone, including city planners, resilience officers, transportation planners, tribal leaders, and residents.

Typing an address or picking a point on a map in the CMRA Assessment Tool reveals climate projections related to extreme heat, drought, flooding, and wildfire. The results show projections for the years 2015–2044, 2045–2064, and 2070–2099 on the basis of two climate change scenarios. In one scenario, global emissions of heat-trapping gases are eliminated by about 2040, and in the other, emissions increase through 2100. Among the many indicators are projections related to extreme rain or heat, number of consecutive dry days, and percentage of coastal counties impacted by global sea level rise.

For example, extreme heat can be judged based on the annual number of days exceeding 105 degrees Fahrenheit. The results may reveal 22–42 days per year by the end of the century compared with

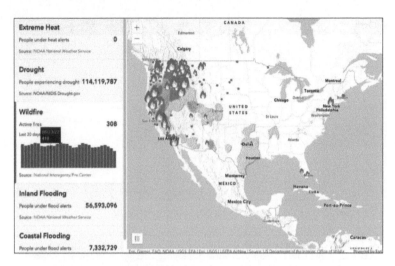

This map, from the CMRA portal, shows active wildfires and areas impacted by wildfires during a 30-day period.

fewer than 4 days in 1990 (the year used as a baseline for emissions calculations by many international organizations). In Washington, DC, the number of days with temperatures exceeding 95 degrees Fahrenheit grows from nearly 19 days per year in the early part of the century to as many as 64 days by the last years of the century.

Although climate action strategies have become commonplace in some larger metropolitan areas, including Boston, Miami, and Los Angeles, a data-driven approach was out of reach for others. The CMRA Portal aims to fill knowledge gaps—helping communities identify climate threats so they can prioritize resilience-building actions and discover programs offering funding to make solutions happen.

For example, a state, local, or tribal leader could use the tool to better understand how temperature, precipitation, and flooding conditions are projected to change locally. The tool could help those leaders generate straightforward hazard reports that inform future

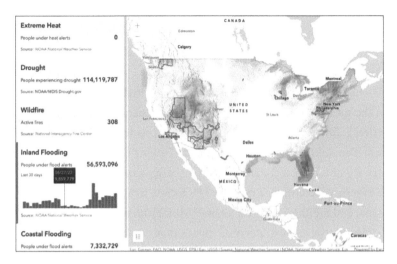

This map, from the CMRA Portal, shows areas impacted by inland flooding during a 30-day period.

projects, a climate action plan, or a data-driven proposal seeking funding.

The tool also shows areas designated as disadvantaged communities—based on an environmental justice score—making them eligible for prioritized funding. The aim is for equity to be a priority when government leaders design and implement resilience measures. The White House's Justice40 initiative ensures that 40 percent of an eligible program's federal funding is spent on communities that are "marginalized, underserved, and overburdened by pollution."

A grant officer can use the tool to review proposed projects by their locations, ensuring that the funds go to projects that are the most needed now and that address a community's resilience for future generations. The grant officer can ensure that funding awards are distributed equitably by verifying priority eligibility funding through the Justice40 initiative.

Local residents also can access the portal to learn more about the climate-related hazards that may impact them, browse maps in the assessment tool, and create a report to share with others to inspire action.

In addition to information about projected climate conditions in the future, the portal highlights links to federal funding resources, federal climate policies, and proven solutions from other communities. The linked US Climate Resilience Toolkit offers videos and stories about what others are doing. One story explains how a tribe in northern Wisconsin has replanted its forests to adapt to future conditions, and another story shares how Minnesota modified its transportation system and extended the hours that cooling centers are open to help people get relief during hot weather.

Technologies such as the CMRA Portal enable dynamic collaboration—the collection and integration of information across agencies to enable better decisions, share solutions, and create a means for supporting evidence-based funding requests for all communities.

The CMRA Portal's collection of relevant climate data also includes open data that can be accessed from the portal and combined with local GIS data or incorporated into assessment tools. The content and open data services invite users to configure new tools and maps that address their local concerns.

"There is no more important service that the federal government can provide to all Americans right now than completely up-to-date information on the climate impacts, on a geographic basis, which are hitting our communities, causing the loss of lives and loss of livelihoods," said David J. Hayes, special assistant to President Joe Biden on climate policy.

A version of this story titled "White House Portal Helps Communities Assess Exposure to Climate Hazards" by Patricia Cummens originally appeared in the *Esri Blog* on September 16, 2022.

ALONG THE MEKONG RIVER, DEVELOPMENT CREATES SUSTAINABILITY CONCERNS

Esri

RAPID GROWTH IN THE NUMBER OF DAMS ALONG THE Mekong River is transforming Southeast Asia's energy, food, transportation, security, and ecological networks. Eleven hydropower dams now span the river before it leaves China, with hundreds more dams planned or under construction in the other countries that contain parts of this vital watershed. The consequences of this construction to people and the environment are still largely unknown.

Scientists at the Stimson Center, a think tank focused on enhancing international peace, have been monitoring the hydroelectric power projects for their impacts on regional stability and the food-water-security nexus.

"Our core message is that nonhydropower renewable energy, like solar and wind, can replace hydropower with far less disruption," said Brian Eyler, director of the Southeast Asia and Energy, Water, and Sustainability programs at Stimson. "Our static map of Mekong mainstream dams is very popular for use in the media and for research presentations, but we realized that a static map didn't show the full picture."

With support from the US Agency for International Development (USAID) and The Asia Foundation, Eyler's team created the Mekong Infrastructure Tracker, a database of visual presentations, spatial analyses, and shared expertise. The dashboard was created using GIS with help from Esri partner Blue Raster.

"In addition to tracking dams, we had the idea to track solar projects after seeing a big uptick in the region in 2018," Eyler said. "Our partners at USAID asked us if we'd track all infrastructure in

Cambodia, Laos, Myanmar, Thailand, and Vietnam. It didn't take much consideration to jump at this opportunity because infrastructure is a hot issue in the Mekong region."

The tracker catalogs projects for power generation and industrial development, as well as road, rail, and waterway transportation. It also includes tools to analyze and quantify the impacts of infrastructure projects.

Construction on the existing dams has displaced thousands of people. Other river people who subsist on fishing and riverbank agriculture are feeling effects of the ecological damage to the ecosystems. In the village of Baan Huay Luek in northern Thailand, for example, villagers speak about how the dams have switched the rules of nature, with dry seasons no longer dry and wet seasons no longer wet.

The Mekong River starts on the Tibetan Plateau in China and flows through Myanmar, Laos, Thailand, Cambodia, and Vietnam before emptying into the South China Sea. The river has been deemed Southeast Asia's most important waterway. It is central to the lives and livelihoods of millions of people and serves as a food source. The Mekong watershed is known as the "rice bowl" of Asia, and 20 percent of the world's freshwater fish catch comes from its waters.

Governments and activists interested in Mekong watershed development have contended with the lack of information or awareness of existing or planned projects.

"This information is usually something that private-sector organizations collect and then distribute only to their clients," said Regan Kwan, research associate at the Stimson Center and manager of the Mekong Infrastructure Tracker project. "Through GIS, we make the data available so that anyone can visualize it, analyze it, and contribute to it."

The Stimson Center works with a range of stakeholders on this and other projects, focusing on community engagement and capacity building. The Mekong Infrastructure Tracker has drawn interest and participation from national development banks, government agencies, ministries, international NGOs, local grassroots organizations, the private sector, academic researchers, and individuals.

"Our Stimson team is relatively small, and we're collecting thousands of data points," Kwan said. "We're bound to make some mistakes, skip things, or not see everything others are seeing. It's great to have others use our data, share their viewpoints, and fill in gaps."

In addition to serving as a singular accurate data source, the Mekong Infrastructure Tracker helps guide foreign policy and assess investment opportunities.

"Having all this information transparently and freely available gives those looking to achieve a more sustainable future a resource to ask questions anytime they want," Eyler said. "No one has provided this depth of information before."

The Stimson Center avoids advocacy, instead allowing partners to draw their own conclusions from the data. One such partner—EarthRights International, based at the Mitharsuu Center in Chiang Mai, Thailand—trains young activists on development issues, environmental impacts, and human rights law and policy. During a seven-month program, called the EarthRights School, students take classes and learn to conduct research.

"We've been encouraging our students to use the Mekong Infrastructure Tracker to map out various projects in their home countries in order to support their advocacy," said William "BJ" Schulte, Mekong policy and legal adviser for EarthRights International. "We value the Tracker as a strong tool that can support networks of activists and community leaders to address the regional energy strategy in the Mekong region as we see a growing reliance on fossil fuels, especially coal."

One of the reasons fossil fuels are gaining interest in Cambodia is that existing dams haven't produced as much energy as predicted, largely because of drought. Coal causes concern because it releases the most greenhouse gas of any fuel, and the hazardous waste by-product, coal ash, is hard to safely store.

"One of our alumni alerted us to a coal-fired power plant that's going to be sited within [Preah Monivong] Bokor National Park," Schulte said. "We're working with local activists to understand the local impacts."

EarthRights International has participated in adding data from communities to the Mekong Infrastructure Tracker to raise awareness about such projects.

"The tracker gives us the ability to draw out where all the projects are and see how they impact each other," said Naing Htoo, Mekong program director at EarthRights International. "We use it to see the regional perspective, including the financiers behind the projects, which includes a lot of cross-boundary investment."

Many government leaders are seeing the Mekong Infrastructure Tracker data and getting a greater regional perspective.

"At first, we had some anxiety about presenting our data to governments in the region, because satellite imagery and other forms of remote sensing data are often seen with suspicion," Eyler said. "We found that data which shows the speed of change in the region and the ease of visualizing and analyzing the data appease concerns. It was the first time they could see what's happening in other countries."

China's Belt and Road Initiative plays a leading role in much of this development. The Mekong Infrastructure Tracker promotes awareness of this and other development programs.

"The data can be used to assess the geopolitics, seeing areas where China, or other external investors from Japan, Korea, or the United States, is more heavily invested," Eyler said. "You can see how many projects are planned and how many are moving forward. Knowing the gaps is useful information."

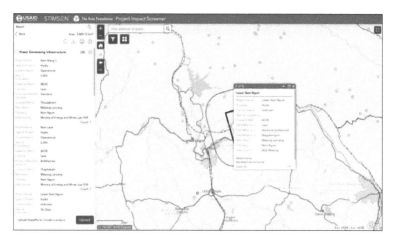

The Project Impact Screener is a tool to identify impacts from infrastructure development in the Greater Mekong subregion.

With more infrastructure projects being planned, and with the rising risks from climate change, EarthRights International workers are also seeing more negative impact on Indigenous people.

"Due to their locations, many of the large infrastructure projects disproportionately impact ethnic minorities and Indigenous peoples," Schulte said. "When they try to raise their voices, they are increasingly attacked or imprisoned, or even disappear. A lot of the fossil-fuel projects are backed by powerful people with significant financial interests."

The transparency of the Mekong Infrastructure Tracker—along with the elements of community science that allow anyone to contribute—encourages participation from a diverse array of people. The Stimson Center takes a data-accurate and research-agnostic approach to the platform to engage participants and encourage dialogue. Administrators are driven to gather complete data and present the truth.

"If anyone using our geodatabase realizes it differs from what they know about an infrastructure project, and they have sources

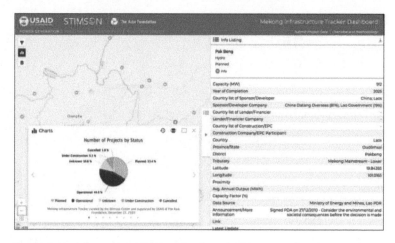

The Mekong Infrastructure Tracker database builds on existing data to present a comprehensive source of information on energy, transportation, and water infrastructure in the Mekong countries, illustrated in this dashboard.

to back it up, we provide an ArcGIS Survey123 form to fill in missing information," Kwan said. "We review it, and if it checks out, we make the update."

The Mekong Infrastructure Tracker has a Facebook group that participants use to drop links to news reports about infrastructure projects. If the news signals a new project, then it's added to the database. Any information about changing timelines or project details are noted. In addition, hackathons are held to fill in data gaps and correct inaccuracies.

The dashboard focuses on five main datasets for energy, transportation, and water infrastructure projects and has data about the surrounding environment as well as people, including ethnic groups. Data on energy projects—including biomass, coal, gas, geothermal, hydro, mixed fossil fuel, nuclear, oil, solar, waste, and wind—is curated along with transportation details on upgrades to canals, urban and high-speed railways, and national roads. The site also covers industrial development at airports, railway stations, and sea or inland ports, as well as special economic zones that provide investment incentives.

"We work to ensure the data is accurate and complete," Eyler said. "The vision is to continue to provide a mix of technical and simple-to-use tools to analyze the data and improve decision-making in the region."

The Mekong Infrastructure Tracker database provides robust environmental and social impact indicator layers that allow anyone to see and study the risks and benefits of various development scenarios.

"A long-term goal is to create a scenario planning application which gives users a chance to generate their own investment scenarios and assess costs and benefits across a variety of indicators," Eyler said. "Users could compare scenarios and even take their scenarios to other policymakers and planners to talk through multiple development pathways and choose the most optimal for their own locality or for the region at-large."

A version of this story titled "Along the Mekong River, Development Creates Sustainability Concerns" by Jen Van Deusen originally appeared in the *Esri Blog* on January 15, 2021.

REGIONAL DATA PLATFORM STRENGTHENS COLLABORATION AND COOPERATION

Harvard Kennedy School

THROUGHOUT HIS CAREER, REX RICHARDSON HAS embodied the public service ethic. His service-oriented career began as student body president at California State University, Dominguez Hills, and then continued as a Service Employees International Union (SEIU) leader. He also advocates more equitable public policies as a Long Beach city council member. In 2022, he was elected as the city's first Black mayor.

Richardson has also served as president of the Southern California Association of Governments (SCAG), the largest—and most successful—council of governments in the country. It serves six counties, 191 cities, and 19 million residents.

SCAG is undertaking one of the most significant and ambitious cross-jurisdictional mapping initiatives in the country, the SCAG Regional Data Platform (RDP). When asked to highlight his primary goals for RDP, Richardson focused not on the technical details, but instead referenced how it will further his commitment to equity, inclusion, and transparency. The RDP website lists these goals for the hub:

- Provide access to data, modern tools, and best practices that support stronger planning and information-based decision-making at all levels.

- Streamline the exchange of data with jurisdictions and partners across the region while establishing procedures and standards for geospatial data consistency.

- Establish a community of planners, GIS professionals, and practitioners to foster collaboration and collective learning, as well as guide the long-term growth and evolution of the RDP.

Richardson explained that "lots of times, positive outcomes seem elusive, and a major reason for that is because policy makers fail to collect and visualize all the relevant data, particularly demographics. Data platforms should be a tool, a solution to help local and regional governments chart a course forward and track progress as well."

With the RDP, SCAG will allow geographers to see how their technical skills are translated into policy breakthroughs and senior officials to see the potential of using ArcGIS software applications to aid the community.

SCAG's Future Communities Initiative, a plan to deploy smart technology and use data analytics to reduce traffic and improve air quality, led to a commitment to build RDP, which will bring together information from members on demographic, economic, land-use, and transportation data. RDP will provide technical resources for in-depth analysis locally and regionally.

Two factors drive SCAG's mapping goals. The first recognizes that technology drives major changes in the region. The second is that data captured using modern technologies grows rapidly and exponentially. The organization's local government members needed assistance harnessing the changing technologies and data to help cities prepare for the future.

Darin Chidsey, SCAG's chief operating officer (COO), helps the organization, which also serves as the region's metropolitan planning organization (MPO), and its members conduct more strategic planning activities. SCAG operates as a regional policy and resource center as cities plan sustainable programs. The RDP initiative serves as

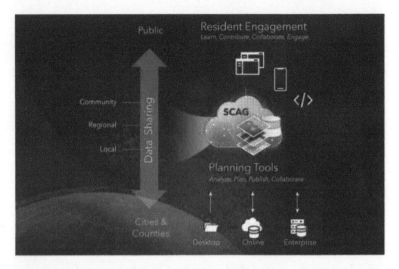

The RDP streamlines data sharing between member cities and SCAG and gives access to the same up-to-date data for planning purposes.

a hub for collaborative data sharing and regional government. "The well-being of the region is ultimately tied to the ability of local jurisdictions to plan for their own futures and share those plans in the form of land use," Chidsey said.

The GIS-supported RDP also furthers goals that include streamlining data sharing, improving transparency and collaboration, and creating equitable outcomes.

With this platform in place, SCAG and local jurisdictions access the same up-to-date data for planning purposes. Chidsey underscored the importance of data standardization and usability of data to support multi- or cross-jurisdictional transportation and city planning where plans and actions in one city can impact outcomes in another. Chidsey likens his team's potential to that of a good baseball team during spring training. The opportunity for participating cities to build on baseline data provided by RDP by continually incorporating other data over time will create value for all member cities.

The map-based planning platform also provides opportunities

for transparency and collaboration between member cities and with SCAG. Richardson explained how RDP will improve collaboration:

Currently, every city in California is responsible for drafting its own general plan. General plans, according to Los Angeles City Planning, "serve as blueprints for the future, describing policy goals and objectives to shape and guide the physical development of the city." Although each city has its own plan, general plans include the individual puzzle pieces that fit together to bring the region into focus.

A major benefit of the initiative is that it facilitates data sharing and collaboration between municipalities developing their general plans as SCAG develops the coordinated Regional Transportation Plan/Sustainable Communities Strategy. Chidsey said the regional plan represents "an aggregation of a lot of data sources—we know our cities have the best data on their area, so this will ease the process of sharing it and increase access to each other's information." Beyond improving ease of access, this platform allows data sharing to flow more freely and continuously with minimal interruption to any city's operations.

Collaboration built on spatially oriented data sharing also increases cooperation in day-to-day governance, not just within cities but between them. When the issue is high-quality transit or good jobs or sustainability policy agendas, politics can interfere in the policy-making process. In those situations, SCAG's joint planning efforts require parties to work with the same data to help overcome differences.

Using shared data increases the chances the parties will communicate more effectively, helping create a symbiotic relationship between regional planning and local planning. Data sharing also benefits individual jurisdictions that can more easily access data from neighboring communities and the region. Cities can plan within the context of a larger, more interconnected system.

Collaboration in government through platforms such as RDP

can occur in ways not possible before, but it's not just about collaborating with government. For SCAG, collaboration between government and residents and community groups is also important. Their voices are sought to inform planning and governance.

Engaging with planners, residents will benefit from the transparency and high-quality visualization provided by the platform. SCAG plans broad use of ArcGIS tools, including map stories, public survey applications, 3D visualization, and digital twin modeling. SCAG developed an effective map story that describes RDP and explains its importance.

"This [platform] will allow us to all have one conversation about the most important issues that face our region, from housing to transportation to jobs. We can focus on these issues from an equity standpoint and collaborate and engage civically around this platform," said Richardson.

Once the benefits of RDP are understood, its connection to addressing equity concerns becomes clear. Sharing data, transparency, and spatial visualizations to enhance a shared narrative will keep a community engaged and help overcome the not-in-my-backyard (NIMBY) effect that occurs in more balkanized and opaque approaches.

Richardson emphasized this valuable aspect. "Patterns of disparity result from aspects of a system that need to be changed, and the only way to shift that system is to present alternative systems based on data. Equitable outcomes are elusive, just like other policy goals, because we often fail to account for data, demographics, and rates of disparity. And so once we have the same common denominator, we can begin to track our outcomes. A data platform and data in general is not the end, it's the means to the end—the language that we'll all need to use in order to actually chart and track a path to a more inclusive society."

GIS tools will provide context to difficult discussions that often involve social factors. SCAG envisions the platform will "allow big data to be the lens or the backbone of the conversation from which we can extract real fact-based insights that further better decision-making amongst multiple stakeholders who need to make decisions based on what they feel is truly best for their own communities."

Although it may seem less glamorous at face value, one of the most transformative features of SCAG's approach is its work assisting member cities in better using technology generally and ArcGIS technology specifically. Its Open Data/Big Data committee focused on how SCAG could provide resources and supporting data tools after it found that nearly 75 percent of members responding acknowledged that they lacked the financial and staff resources to support data and technology projects.

SCAG's role includes facilitating use of the regional data platform, including helping members develop the data and analytic capacity necessary for insights. These services will help level the playing field for all member agencies, regardless of their current capacity.

As members use the tools, they can understand their effectiveness and increasingly apply them in more and deeper ways. This cycle of RDP and ArcGIS software usage will drive substantial public value. Increased value will come from access to the tools and grants made to members and by granting Esri licenses to assist local governments in planning that supports transit, housing, and economic opportunities.

SCAG has developed a track record of providing technical services starting with GIS training, analyses, and hardware to cities that lack necessary capabilities. SCAG's nationally significant work illustrates that for governments to be successful in serving residents, they must plan and operate recognizing the importance of place.

A version of this story titled "Regional Data Platform Strengthens Collaboration and Cooperation" by Stephen Goldsmith originally appeared in the 2021 Winter issue of ArcUser.

SINGAPORE BUILDS A MASSIVE MARITIME SPATIAL ATLAS TO MANAGE CLIMATE CHANGE

Esri

WHAT THE FUTURE HOLDS FOR SINGAPORE'S COASTAL areas is a matter of strategic global importance and local existential survival. The Port of Singapore, located where the Indian Ocean meets the Pacific, is the world's largest and busiest transshipment point. Much of the island itself lies perilously just a few meters above sea level, making it vulnerable to the effects of climate change and rising sea levels.

Until recently, technological limitations have hampered abilities to map the ocean depths and address challenges such as those faced by Singapore. But since 2019, the Maritime and Port Authority (MPA) of Singapore has managed GeoSpace-Sea, a national marine spatial data infrastructure of Singapore's coast and coastal waters. GeoSpace-Sea aims to bring clarity to this cartographic frontier, using GIS to assemble and display layers of maritime and marine data.

"The sea is a planet that has not yet been fully explored," said Parry Oei, a hydrography adviser for MPA. "It's quite mind-boggling how everything is linked to everything else."

Oei is one of the prime movers behind the creation of GeoSpace-Sea. For scientists, GeoSpace-Sea is a way to understand how the world's oceans are changing, and how these changes are impacting Singapore. For planners, GeoSpace-Sea is a way to anticipate these changes and make predictive analysis about how they will affect Singapore using a shared repository of information. For port officials, the higher-resolution data from the project will underpin operations, allowing them to see and consider the consequences of decisions, such as the impact of dredging on sea life.

Oei traces GeoSpace-Sea's provenance to the late 1980s, when he was working in Hobart, Australia, as part of the Asia-Australia Project, a collaborative effort to understand the emerging signs of climate change.

It was during this time in Australia that Oei began to think about mapping the complexity of the ocean. Climate change would surely impact the oceans, and the oceans would in turn affect climate change.

The question was how to gather data—and what to do with it. At that time, GIS was a niche technology used primarily by trained GIS experts. As this group did not generally include hydrographers, Oei found little enthusiasm for this use of GIS within Singapore's government.

"Hydrographers are always the supporting actors," he explained. 'At the end of the movie when the credits roll, we're the crew list. Sometimes it goes so fast you don't even see your name."

In the ensuing years, GIS became a cloud-based technology, intuitive for cross-discipline collaborations and capable of including spatial data at all scales and without limits on data volumes. By the mid-2000s, as climate change became more of a pressing concern, and sea level rise posed an existential threat to Singapore, Oei's idea gained momentum.

"Climate change gave us the profile and attention we needed," Oei said.

Sea level rise is not the only climate change variable Singapore will experience. Intense storms will increase, causing storm surge and inland flooding.

As a marine geospatial knowledge repository, GeoSpace-Sea could increase understanding of extreme weather events. For instance,

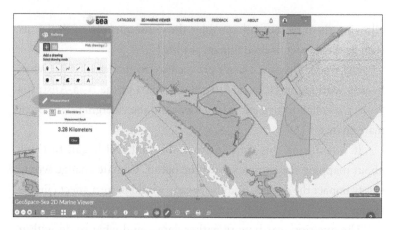

Viewing the coastline using the GeoSpace-Sea 2D Marine Viewer gives a detailed view and tools to mark up and measure on the map. Screenshot courtesy of the Maritime and Port Authority of Singapore.

until recently it was assumed that water flow in the Singapore Strait was chiefly affected by the Northwest Monsoon, a period in winter and early spring when the Indian Ocean's trapped heat lures cold air from the Himalayas. Scientists now believe the Southwest Monsoon—the summer season when winds blow moist air from the ocean toward the Himalayas—wields a stronger influence.

As GeoSpace-Sea expands, Oei hopes to add more weather data, to further this integrated approach. The MPA also plans to make some of GeoSpace-Sea's layers available to the public once sufficient cybersecurity protocols are established.

"The question will be how do we get rid of water when it floods and how do we prevent seawater from coming in," Oei said. "I have no easy answers, but I do believe we need to monitor everything in the environment—weather patterns, temperature, sea level rise—because it's all part of climate change."

For now, GeoSpace-Sea is primarily useful as an unfolding historical document that lets researchers study past trends to predict future developments. New data is added as it becomes available.

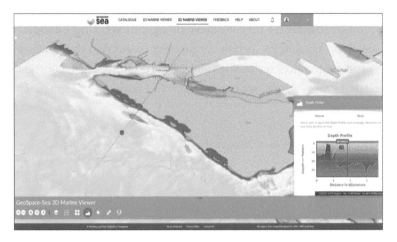

The GeoSpace-Sea 3D Marine Viewer lets users see below the surface with details of what lies below. Screenshot courtesy of the Maritime and Port Authority of Singapore.

GeoSpace-Sea will increase the use of sensors and the IoT to have some of the data update in near real time. "With modeling, there's only so much you can do," Oei said. "But if you integrate modern sensors, this truly becomes a digital twin, and you become more proactive than reactive."

The ocean is an immensely complicated system. As GeoSpace-Sea grows, the ability of scientists to accumulate knowledge will increase their understanding of this complexity.

GeoSpace-Sea also will introduce an AI aspect, using machine-learning protocols to search for patterns and hot spots. As an example, Oei suggests that GeoSpace-Sea could help Singapore better understand the conditions that lead to red tide, when algal blooms wreak havoc on its coastal ecosystems.

"What makes AI so interesting is that it could uncover things we don't know, but we honestly can't say what it will show us," Oei said.

The long-term goal of GeoSpace-Sea, he explained, is to further scientific progress by gathering as much information as possible to address questions we may not have thought of yet. Understanding something as complex as how Singapore's connection to the ocean is evolving requires technological sophistication, but also constant collaboration. "Alone, we can go fast," Oei said. "But together, we can go far."

A version of this story titled "Singapore Builds a Massive Maritime Spatial Atlas to Manage Climate Change" by Chris Power originally appeared in the *Esri Blog* on July 7, 2022.

AFRICA GEOPORTAL BRINGS TOGETHER A CONTINENT OF GIS USERS

Esri

G IS USERS WHO LIVE IN OR WORK ON TOPICS RELATED TO Africa now have a community geospatial platform for the continent: Africa GeoPortal.

The portal brings together geospatial data, geospatial tools, and learning about GIS, said Matthew Pennells, former Africa manager and current director of global community engagement for Esri in Dubai. "All this is free for single users to access if they're working in Africa or on African-based geospatial challenges. It's also designed to be helpful regardless of how much experience in geospatial or GIS technology a user has."

Before Africa GeoPortal launched in early 2019, there wasn't anything that connected the geospatial community in Africa, despite many common challenges, he said.

"How agricultural problems are solved in Kenya and Ghana is similar, yet how agricultural problems are solved within Ghana and the United States is very different," Pennells said. "So, there's value in having this as a continental platform."

Africa GeoPortal comes with content from ArcGIS Living Atlas of the World that's curated to relate directly to Africa. Users also get access to ArcGIS Online; apps and analytical tools such as ArcGIS Collector, ArcGIS Survey123, ArcGIS StoryMaps, and ArcGIS Insights℠; and Esri's e-learning materials, including videos and interactive web courses that teach portal concepts and capabilities.

But the value of Africa GeoPortal derives from community-contributed data.

"People bring in data they might collect at a local or national or continental level, especially if they need a place to store it," Pennells said. "We also have governments adding data, so statistical agencies and such that want to expand access to their open data. And then we have broader partners in the geospatial industry, such as the Earth observation initiative Digital Earth Africa."

The idea behind Africa GeoPortal is for the community of diverse users to inspire each other and build on each other's work.

"If someone is collecting data on health center locations in Kenya and creates a map to show people where the newest one is, we want other people to see that and take it further, maybe by doing that in Ghana," he said. "So it's a combination of people contributing data and Esri giving them space to share their data and inspire other users."

Once users sign up for a complimentary account, they can create and store data, make maps, and build solutions. For data scientists, the platform has a range of notebook tools. If users want to learn more about any of this, Africa GeoPortal points them to specific e-learning materials—many of which use data, examples, and scenarios from throughout the continent—that they can use to build their skills.

For instance, building an app from data about health facilities in Kenya no longer requires the expertise of an app developer. The data and tools needed are available in Africa GeoPortal. The user can find and download the data, locate the appropriate geospatial tools—such as Survey123 or Collector—to fill in any missing data, and bring that information into a simple app for sharing.

Africa GeoPortal highlights themes that affect most, if not all, of the continent. Resources for responding to the COVID-19 pandemic, for example, were highlighted for most of 2020. Navigating to the subpage, users found official dashboards containing COVID-19 data for African Union member states and for countries ranging from Morocco to South Africa. It also hosted reliable data and tools from independent users, such as a dashboard from a user in Ghana.

Another subsection of the home page spotlighted data about the desert locust crisis afflicting the Horn of Africa. It pointed users primarily to an open data hub built by the UN Food and Agriculture Organization (FAO) and other datasets on locust swarms, control operations, hot spots, and more.

"What we really want to see for these subtopics is for people to go on and build localized solutions—whether that be local to their geographic problem or local to their particular thematic area—that are based on the authoritative data assembled on the Africa GeoPortal," Pennells said.

The site is further broken down into country and organizational pages, where data, examples, and stories are arranged around specific needs and objectives. Anything from these pages is accessible on the wider Africa GeoPortal, and users within the more streamlined geoportals can still access everything there as well.

The National Office of Technical Studies and Development, known by its French acronym BNETD, maintains a country-level geoportal for Côte d'Ivoire in West Africa to make its open data more accessible.

"Right now, we are just publishing data for everyone," said Fernand Balé, director of the Geographic and Digital Information Center (known as CIGN) at BNETD. "But we are building a network of users in every field—in social, economic, and government fields."

Sharing data is just the first step in what BNETD seeks to do.

"We want to build apps in order to help our government and decision-makers make good decisions," said Balé. "In our vision, people could share very smart datasets—accurate data—in order to plan infrastructure or access to education and health centers, to achieve more in the private sector, and so on. We want to build a very powerful geodatabase for our users. We want to show how geographic tools can help improve government programs and people's lives."

But first, users need access to accurate data—and a lot of it—which, according to Balé, can be hard to find.

"My big picture [vision for our use of Africa GeoPortal] is to be a reference—the place where everyone, every government organization, can find what they are looking for to improve their work, to improve people's lives, to fight poverty, to fight climate change, and so on," he said.

Another organization working to disseminate important geospatial data through Africa GeoPortal is Digital Earth Africa, which aspires to make free and open satellite data of the entire continent available in analysis-ready formats.

"Esri is the GIS package that is most abundantly used, especially in Africa," said Aditya Agrawal, former senior program adviser for Digital Earth Africa. "Being able to have our data available within the Africa GeoPortal enables that data to be accessible by more users."

The team at Digital Earth Africa also cares about how its data is being used, which is something it can see through Africa GeoPortal. "Being able to understand how [geoportal] users are using the data and what outcomes and stories come out of that is going to be the most important part for us and our community."

One area that Digital Earth Africa is currently focused on is food security. It has tailored some of its satellite imagery so users can visualize land-use change over time.

"We're developing a crop mask for the continent to make... changes in agricultural areas easier to understand at national and local levels," Agrawal said. "All the data can be explored via a map viewer, and eventually, we will develop an ecosystem approach where you can access fit-for-purpose apps."

The organization has also used Africa GeoPortal to release water-related Earth observation data.

"Often, villagers have to go to the nearest water body to fill up their pots with drinkable water. Sometimes, they have to walk far distances to find that, and once they get there, they don't know if there will be water," he said. The goal of the imagery and an associated app is to simplify the process so that people living in these villages can get the water they need, Agrawal said.

Africa Geo-Portal continues to draw more use after its launch in early 2019.

"We want to remove the barriers to entry for working with geospatial data that currently exist throughout much of Africa and make it so that people can direct their energy toward building local solutions that help solve local problems," Pennells said.

A version of this story titled "The Africa GeoPortal Brings Together a Whole Continent of GIS Users" originally appeared in the 2020 Fall issue of ArcNews.

NEXT STEPS

I N A WORLD OF CONSTANT FLUX, DATA IS MORE OPENLY accessible today than ever before. Having an integrated geospatial infrastructure in place provides a new, more powerful way to handle and use open data, giving users the capability to support geospatial thinking and knowledge and a practical way to act collectively and urgently on today's shared challenges.

In this book, you learned how organizations, national and international governments and companies, local and regional governments, and nonprofits use their geospatial infrastructure to connect their communities and broaden efforts to improve collaboration, achieve common goals, and work effectively beyond the boundaries of jurisdictions and restrictive data systems.

If you are new to GIS or considering implementing an integrated, geospatial infrastructure, your next step should be to take a course (or have your relevant staff take a course) that presents the fundamental concepts and workflows needed to plan and create your own collaborative geospatial program within your community.

Exploring the capabilities of geospatial infrastructure is a good place to start.

GIS for Collaborative Communities: An ArcGIS Tutorial Collection

In the GIS for Collaborative Communities: An ArcGIS Tutorial Collection, you will learn how to use ArcGIS software to support the four pillars of an integrated geospatial infrastructure: collaborative governance, interconnected data and technology, community engagement, and geospatial capacity for stakeholders.

You will learn how to

- understand global initiatives supporting your journey,

- explore fundamental concepts and recommended patterns and practices,

- guide your data consumers and equip your providers with geospatial tools,

- develop your workforce to successfully use and share geospatial data,

- collect, manage, and curate essential geospatial data,

- prepare for distribution and sharing, and

- build and nuture your collaborative network of partners, educators, and students.

GIS for Collaborative Communities: An ArcGIS Tutorial Collection is an all-in-one, online course.

Alternatively, you may want to focus your learning on specific ArcGIS software mentioned in the case studies.

Learn by doing

Esri also provides free, hands-on tutorials for people of different experience levels and various roles, such as new users, students, schoolteachers and university professors, data scientists, and GIS professionals.

Since the main theme of this book is using geospatial infrastructure to facilitate collaboration, the following ArcGIS tutorials are recommended as starting points:

Share Maps and Collaborate with Colleagues Using ArcGIS Online

In this tutorial, you will learn how to

- share maps,
- create web apps,
- use groups effectively,
- manage data, and
- set essential configurations for highly scalable web apps.

Essential Guides for OneMap Administrators

In this tutorial, you will learn about

- connecting communities with modern GIS integrated geospatial infrastructure patterns in practice,
- configuring ArcGIS Online and ArcGIS Hub for sharing and collaboration,
- applying good practices for authoritative data providers,
- sharing and collaborating using OneMap, and
- using the OneMap Hub template.

If you like this book and want to see more industry-focused examples of organizations using their geospatial infrastructure to solve problems and work better together, check out the Applying GIS series of books and browse other books from Esri Press at esri.com /en-us/esri-press.

Learn more about these tutorials and other GIS resources for geospatial infrastructure and collaboration by visiting the webpage for this book:

go.esri.com/wbb-resources

CONTRIBUTORS

Brian Boulmay
Richard Budden
Greg Bunce
Patricia Cummens
Mark Cygan
Chris Fowler
Stephen Goldsmith
Brent Jones
Marianna Kantor
James Miller
Chris North
Nick O'Day
Eva Pereira
Linda Peters
Jen Van Deusen

ABOUT ESRI PRESS

ESRI PRESS IS AN AMERICAN BOOK PUBLISHER AND PART OF Esri, the global leader in geographic information system (GIS) software, location intelligence, and mapping. Since 1969, Esri has supported customers with geographic science and geospatial analytics, what we call The Science of Where®. We take a geographic approach to problem-solving, brought to life by modern GIS technology, and are committed to using science and technology to build a sustainable world.

At Esri Press, our mission is to inform, inspire, and teach professionals, students, educators, and the public about GIS by developing print and digital publications. Our goal is to increase the adoption of ArcGIS and to support the vision and brand of Esri. We strive to be the leader in publishing great GIS books, and we are dedicated to improving the work and lives of our global community of users, authors, and colleagues.

Acquisitions

Stacy Krieg

Claudia Naber

Alycia Tornetta

Craig Carpenter

Jenefer Shute

Editorial

Carolyn Schatz

Mark Henry

David Oberman

Production

Monica McGregor

Victoria Roberts

Sales & Marketing

Eric Kettunen

Sasha Gallardo

Beth Bauler

Contributors

Christian Harder

Matt Artz

Keith Mann

Business

Catherine Ortiz

Jon Carter

Jason Childs

Related titles

Getting to Know Web GIS, fifth edition

Pinde Fu

9781589487277

Top 20 Essential Skills for ArcGIS Pro

Bonnie Shrewsbury and Barry Waite

9781589487505

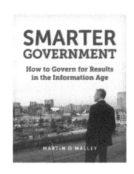

Smarter Government: How to Govern for Results in the Information Age

Martin O'Malley

9781589485242

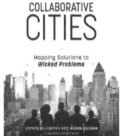

Collaborative Cities: Mapping Solutions to Wicked Problems

Stephen Goldsmith & Kate Markin Coleman

9781589485396

For information on Esri Press books, e-books, and resources, visit our website at

esripress.com.